Op-Amp Circuits
Simulations and Experiments

by Sid Antoch

ZAP Studio, LLC · · · Philomath, Oregon

ISBN: 978-1-935422-15-0

Published by:

ZAP Studio, LLC
PO Box 988
Philomath, OR 97370
www.zapstudio.com

Contents

Introduction

Operational amplifiers have a very broad range of application. This book focuses on the fundamentals which are applicable to many applications. All of the simulations and experiments demonstrate basic operational amplifier principles. The experiments may be easily modified and may serve as the basis for other applications.

This book may be used as a circuit design and application reference for hobbyists, experimenters, and students. It may also be used as a supplement to a college level operational amplifier course and laboratory. An understanding of electric circuit analysis, semiconductor devices, and college level algebra are pre-requisites for this book.

Simulation examples are presented using *LTspice*, a simulation program available as a free download from *Linear Technology*. *TINA-TI*, a simulation program available as a free download from *Texas Instruments*, is also introduced.

Experiments provided may be performed using a solder-less breadboard, inexpensive parts, a small power supply, and a digital or USB oscilloscope. Some experiments also require a function generator. The circuits are provided in their basic and simplest forms. The experimenter may modify and augment the circuits as needed for particular applications.

Chapter 1: Op-Amp Characteristics

An operational amplifier is a device with a very high voltage gain, very high input impedance, and very low output impedance. It is a versatile device that can function as an amplifier, oscillator, filter, integrator, differentiator, or comparator.

Op-amps were originally designed with vacuum tubes in the 1940's. Transistors replaced the vacuum tubes in the 1950's and by the 1970's op-amps were available as integrated circuits.

Operational Amplifier: Open-Loop DC Model

Figure 1-1 shows a simple model of an operational amplifier. The V+ terminal is called the "non-inverting input" and the V- terminal is called the "inverting input". E is a voltage-controlled voltage source whose output voltage, Vo, is given by:

$$Vo = A\left(V_+ - V_-\right).$$

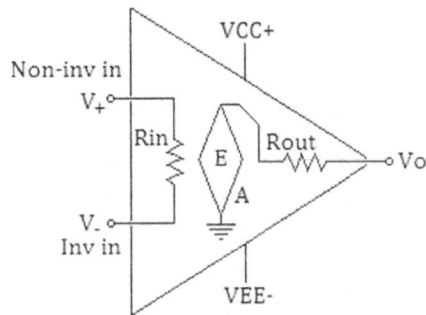

Figure 1-1

Any voltage dropped across Rout is negligible.

In other words, the output voltage is equal to the voltage across the two inputs multiplied by the voltage gain A. Typically the value of A is between 10^5 and 10^7. It will be shown later that the value of A decreases as the input frequency increases.

The op-amp's input resistance is represented by Rin. Its open-loop value is typically several megohms. It will be shown later that its value increases considerably in a "closed-loop circuit" (with feedback).

The op-amp's output resistance is represented by Rout. Its open-loop value is typically less than 100 ohms. It will be shown later that its value decreases considerably in a closed-loop circuit.

VCC and VEE are the op-amp's power supply voltages.

The model in figure 1-1 will be used to derive the following characteristics of op-amp: closed-loop voltage gain, closed-loop input impedance, and closed-loop output impedance. Also, in a typical closed-loop circuit, the voltages V+ and V- are very nearly equal.

1

Non-Inverting Amplifier

Figure 1-2 shows the configuration for a non-inverting amplifier. Rf and Ri provide negative feedback. Rs is the load resistance for the signal source, Vs.

Non-Inverting Amplifier Closed Loop Voltage Gain

Rout is much smaller than Rf and it will be assumed to be zero. The application of ohm's law and algebra result in the following equations:

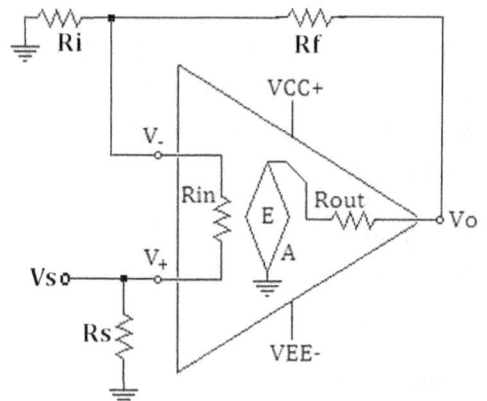

Figure 1-2

$$Vo = A(Vs - V_-). \qquad V_- = \frac{Ri}{Ri + Rf} Vo. \qquad Let \; \beta = \frac{Ri}{Ri + Rf}.$$

$$Vo = A[Vs - \beta Vo] = AVs - \beta AVo. \qquad Vo[1 + \beta A] = AVs.$$

$$\frac{Vo}{Vs} = \frac{A}{1 + \beta A} = \frac{1}{\frac{1}{A} + \beta}. \qquad 10^5 < A < 10^7 \; therefore \; \frac{1}{A} \approx 0.$$

Non-Inverting amplifier voltage amplification is: $A_{NI} = \dfrac{Vo}{Vs} = \dfrac{Ri + Rf}{Ri} = \dfrac{1}{\beta}.$

β is called the "feedback factor", the fraction of the output voltage, Vo, that is fed back to the op-amp's inverting input.

It can be shown that the voltages on the two inputs V- and V+ must be very close to equal as follows:

$$Vo = A(Vs - V_-), \; therefore \; (Vs - V_-) = \frac{Vo}{A} \approx 0 \; (because: 10^5 < A < 10^7).$$

Note that in this circuit V+ = Vs. The maximum value of Vo is the value of the power supply voltage, which typically may range between 5 volts and ±15 volts. For most op-amps, the actual maximum value of Vo will be one to three volts less than the power supply voltage. Exceptions are "rail to rail" op-amps whose maximum value of Vo is very close to the power supply voltage.

Example: If an op-amp's Vo can change from -15V to +15V, and its open-loop gain is 10^5, the maximum difference between V+ and V- is only 0.3mV.

Non-Inverting Op-Amp Input Impedance

The input impedance into the non-inverting input is:

$$i_{in} = \frac{V_+ - V_-}{Rin} = \frac{Vs - \beta Vo}{Rin}. \qquad \frac{Vo}{Vs} = \frac{A}{1+A\beta}. \qquad Vo = \frac{A}{1+A\beta} Vs.$$

Substitute Vo into equaton for i_{in}:

$$i_{in} = \frac{1}{Rin}\left[Vs - \beta Vs\left(\frac{A}{1+A\beta}\right)\right] = \frac{Vs}{Rin}\left[1 - \frac{A\beta}{1+A\beta}\right].$$

$$Zin = \frac{Vs}{i_{in}} = Rin(1+A\beta) \cong (A\beta)Rin.$$

Zin is typically greater than 10^9 ohms because Aβ is typically greater than 10^3 and Rin is typically greater than 10^6 ohms. This high input impedance is in parallel with Rs so the amplifier's input impedance is essentially determined by the value of Rs.

Op-Amp Output Impedance

The output impedance is calculated using the "open circuit voltage over the short circuit current" method:

$$i_{sc} = \frac{AVs - Vo}{Rout} = \frac{AVs - 0}{Rout} = \frac{AVs}{Rout}.$$

Substitute $Vs = Vo\left(\frac{1+A\beta}{A}\right)$ into equation for i_{sc}: $i_{sc} = \frac{Vo}{Rout}(1+A\beta).$

Removing the short circuit at Vo makes $V_{open} = Vo$.

$$Zo = \frac{V_{open}}{i_{short}} = \frac{V_{open}}{\frac{Vo}{Rout}(1+A\beta)} = \frac{Rout}{1+A\beta} \cong \frac{Rout}{A\beta} \qquad (Vo = V_{open} \text{ here}).$$

It can be concluded that the op-amp's input impedance is essentially infinite and output impedance is essentially zero. The op-amp's input and output impedances are insignificant compared to the values of the external components connected to it. This is also true for the inverting amplifier circuit.

Example: An op-amp's open-loop output impedance, Rout, is 75 ohms and open-loop gain, A = 10^6. It is connected as an inverting amplifier with a gain of 100.

$$Zo = \frac{Rout}{A\beta} = \frac{75}{10^6 10^{-2}} = .0075 \, ohms.$$

3

Figure 1-3 summarizes the properties of an "ideal" non-inverting op-amp circuit.

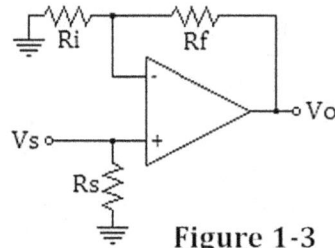

Summary:

Non - Inverting Amplifier

$$A = \frac{Vo}{Vs} = \frac{Ri+Rf}{Ri}$$

$$V_- = V_+ \quad Zs = Rs \quad Zo = 0$$

Figure 1-3

Inverting Amplifier

Figure 1-4 shows a diagram of an inverting op-amp circuit. Rf and Ri form a voltage divider to provide negative feedback. Ri is also the load resistance for the signal source, Vs. Rout, the output impedance op-amp, is assumed to be zero.

The op-amp's input currents are much smaller than the circuit's external currents and are assumed to be zero.

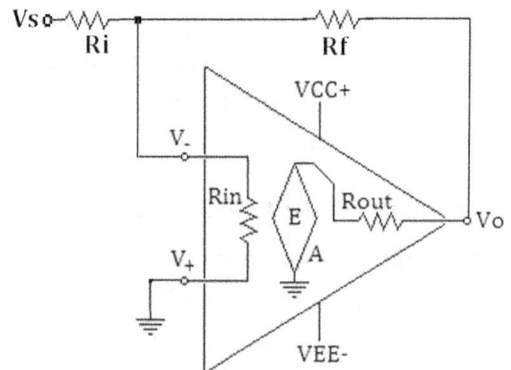

Figure 1-4

The voltage V- is very close to 0 volts because $V_+ = 0$. The V- terminal is called a "virtual ground". The amplification of the inverting op-amp circuit is derived below:

Node voltage eq:
$$\frac{V_- - Vs}{Ri} + \frac{V_- - Vo}{Rf} = 0.$$

Voltage gain eq: $Vo = A(0 - V_-) = -AV_- \Rightarrow V_- = \frac{-Vo}{A}.$

Solve the node voltage equation by substituting for V:

$$\frac{\frac{-Vo}{A} - Vs}{Ri} = \frac{\frac{Vo}{A} + Vo}{Rf} \Rightarrow \frac{Rf}{Ri} = \frac{\frac{Vo}{A} + Vo}{\frac{-Vo}{A} - Vs} \cong \frac{-Vo}{Vs}.$$

Inverting Amplification is: $A_{INV} = -\frac{Vo}{Vs} = -\frac{Rf}{Ri} \quad \left(\text{given} \frac{Vo}{A} \cong 0 \right).$

4

Since the V- input is essentially zero volts, it is considered a virtual ground. The input impedance of the inverting amplifier circuit in figure 1-5 is equal to Ri.

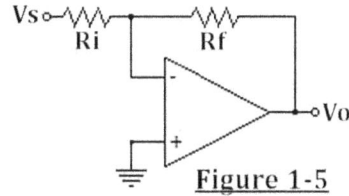

Summary : Inverting Amplifier

$$A = \frac{Vo}{Vs} = \frac{-Rf}{Ri}$$

$$V_- = V_+ = 0 \quad Zs = Ri \quad Zo = 0$$

Figure 1-5

Differential Amplifier

The differential amplifier circuit in figure 1-6 uses both inverting and non-inverting inputs. The amplifications for the inverting and non-inverting inputs, Vs- and Vs+, are given below.

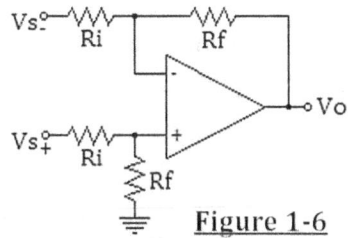

Figure 1-6

$$A_{INV} = -\frac{Vo}{Vs_-} = -\frac{Rf}{Ri}.$$

$$A_{NI} = \frac{Vo}{Vs_+} = \left(\frac{Rf}{Ri+Rf}\right)\left(\frac{Ri+Rf}{Ri}\right) = \frac{Rf}{Ri}.$$

A signal source, V_{diff}, is normally connected between Vs- and Vs+, as shown in figure 1-7. The amplification value of this connection is the differential mode gain, A_{diff}.

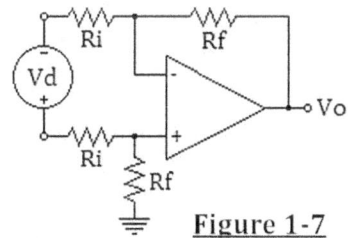

Figure 1-7

$$Vo = (Vs_+ - Vs_-)\frac{Rf}{Ri} = V_{diff}\frac{Rf}{Ri} \quad \Rightarrow \quad A_{diff} = \frac{Vo}{V_{diff}} = \frac{Rf}{Ri}.$$

A signal applied equally to both inputs, as shown in figure 1-8, is called a "common mode" signal. This connection would not normally produce an output, but because op-amps are not perfect, some output will be present. The ratio of this output to the common mode input is called the common mode gain, A_{CM}.

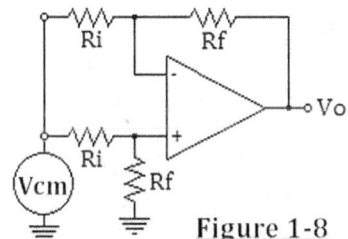

Figure 1-8

An op-amps ability to reject the common mode signal is called its "common mode rejection ratio", CMRR.

An op-amp's CMRR decreases with frequency as shown for the LT1013 on the right. The CMRR for a differential amplifier circuit depends mostly on the precision of the resistors, Ri and Rf. The CMRR of an op-amp itself is usually much better than that of the differential amplifier circuit. If CMRR is critical in an application, very high precision resistors and/or a balancing circuit needs to be used.

LT1013 CMMR vs Frequency

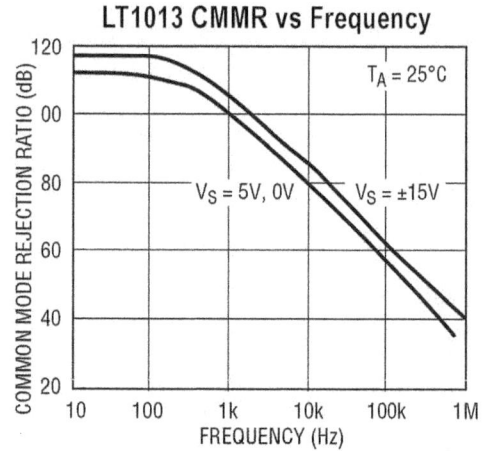

The differential amplifier circuit described here has two disadvantages: its input impedance is relatively low and its inverting and non-inverting input impedances are not equal. These disadvantages are overcome by the "instrumentation amplifier" circuit described in the next section.

Summary : Differential Amplifier

$$CMRR = \frac{A_{CM}}{A_{diff}} \qquad CMRR_{DB} = 20\log\frac{A_{CM}}{A_{diff}} dB$$

$$A_{diff} = V_{diff}\frac{Rf}{Ri} \qquad Z_{IN+} = Ri + Rf \qquad Z_{IN-} = Ri$$

Instrumentation Amplifier

Figure 1-9 shows the basic circuit configuration of a three op-amp differential amplifier. This circuit is commonly referred to as an "instrumentation amplifier".

It has balanced inputs and accommodates a very wide range of input impedance. Instrumentation amplifiers may be designed with individual op-amps or specialty instrumentation amplifier ICs may be used.

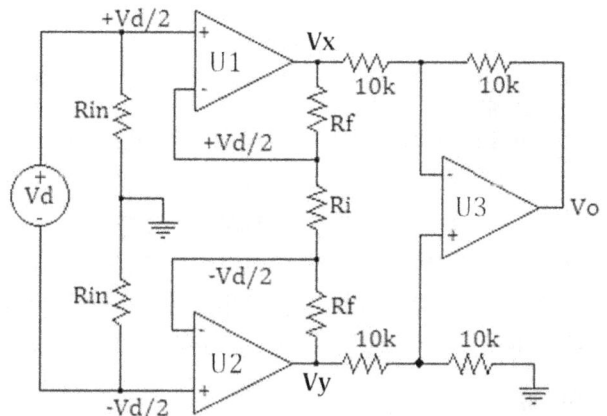

Figure 1-9

6

Refer to figure 1-9 for derivation of the differential amplifier gain presented here. The output of the first stage, U1 and U2, is the differential signal, Vx and Vy. The second stage, U3, is a unity gain single op-amp differential amplifier. The gain of the second stage does not need to be unity.

The differential voltage, Vd, is referenced to ground as +Vd/2 on U1's inverting and non-inverting inputs and as −Vd/2 on U2's inverting and non-inverting inputs. The voltage across Ri is equal to Vd. Ohms law is used to calculate the current through Ri, which also flows through both Rf's. This current is used to calculate the voltages Vx and Vy.

$$I_x = \frac{Vd}{Ri}. \qquad Vx = I_x Rf + \frac{Vd}{2}. \qquad Vy = -I_x Rf - \frac{Vd}{2}.$$

$$Vx - Vy = \left(I_x Rf + \frac{Vd}{2}\right) - \left(-I_x Rf - \frac{Vd}{2}\right) = 2\frac{Vd}{Ri}Rf + Vd = Vd\left(\frac{2Rf}{Ri} + 1\right).$$

$$A_D = \frac{Vx - Vy}{Vd} = \frac{2Rf}{Ri} + 1.$$

The differential amplifier U3 converts the differential output of U1 and U2 to a "single ended" output Vo. Since the gain of U3 is unity, Vo = Vx − Vy.

Summary : Instrumentation Amplifier

$$A_D = \frac{2Rf}{Ri} + 1 \qquad Zin_D = 2Rin$$

Op-Amp Input Offset Voltage and Input Bias Currents

In the op-amp model in figure 1-10, input resistance, Rin, is replaced by two current sources, Ib₋ and Ib₊. These represent the op-amp's bias currents. Vos represents the op-amp's internal offset voltage.

This circuit can represent a non-inverting amplifier by setting Vs₋ to zero, or an inverting amplifier by setting Vs₊ to zero.

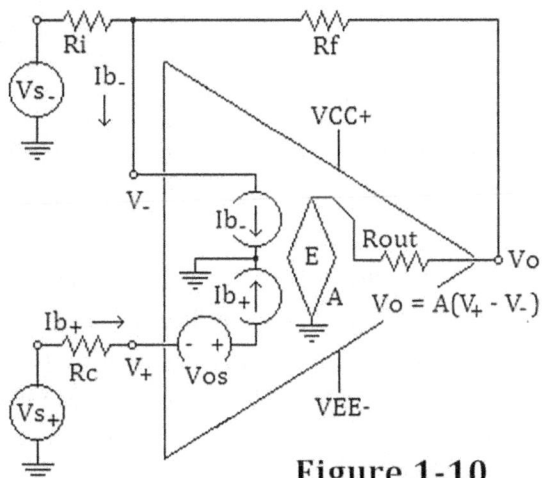

Figure 1-10

7

Inverting and Non-Inverting Amplifier Offset Current Error

Vs.= 0. Vs.= 0. Vos = 0. Voe = output error voltage.

$$Voe=(-RcIb_+)\left(\frac{Ri+Rf}{Ri}\right)+\left(\frac{RiRf}{Ri+Rf}Ib_-\right)\left(\frac{Ri+Rf}{Ri}\right).$$

$$Voe=\left(-RcIb_+ + \left(\frac{RiRf}{Ri+Rf}\right)Ib_-\right)\left(\frac{Ri+Rf}{Ri}\right).$$

$$\text{If } Rc=\frac{RiRf}{Ri+Rf}\text{ then:} Voe=Rc(Ib_- - Ib_+)\left(\frac{Ri+Rf}{Ri}\right).$$

Rc is the bias current compensation resistance. An op-amp with equal bias currents but unequal net resistances to ground will develop an offset error voltage. The correct value of Rc insures that any bias current error is due only to the op-amp's offset current.

$$\text{If } Ib_- = Ib_+ \text{ and } Rc = \frac{RiRf}{Ri+Rf}, \text{ then } Voe=0.$$

Inverting and Non-Inverting Amplifier Offset Voltage Error

Vs.= 0, Vs.= 0.
$$Voe=Vos\left(\frac{Ri+Rf}{Ri}\right).$$

Offset Compensation

Some op-amps have a provision for offset compensation, such as the LM741. Two compensation circuits are shown in on the right in figure 1-11.

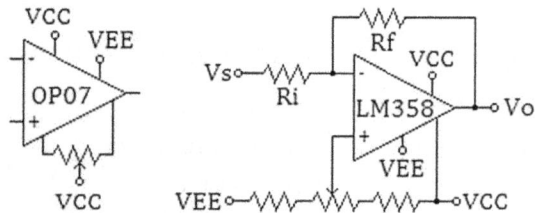

Figure 1-11

The LM358 circuit uses a voltage divider to set the voltage on the op-amp's non-inverting input.

Offset compensation circuits are usually only used for critical applications. Offset error can usually be minimized by the proper choice of op-amps and circuit designs. Cascaded DC amplifers may have an offset compensation circuit in the first stage because the offset error is multiplied by the gain od the amplifiers.

Frequency Response

Figure 1-13 on the right shows the open loop frequency response of the LM358 op-amp on a log-log graph (Bode plot). This plot has the characteristics of a single pole low pass filter with a cut-off frequency of about 8Hz.

The op-amp's gain drops to unity (0dB) at 1MHz. This is called the unity gain frequency. Data books often refer to the unity gain frequency as the "gain-bandwidth-product", GBP, of the op-amp.

Open Loop Frequency Response

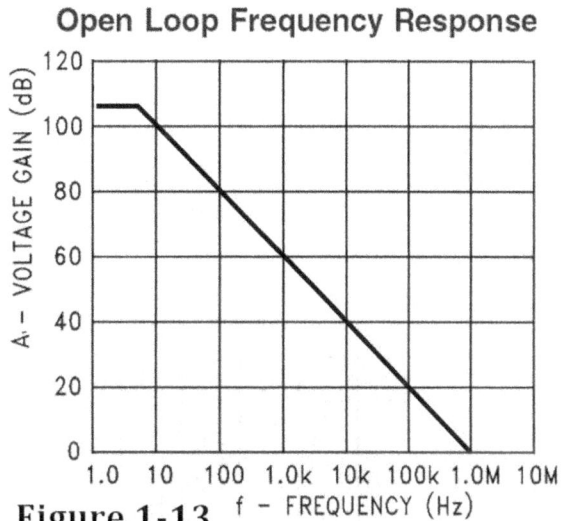

Figure 1-13

An op-amp's GBP extends from DC to its unity gain frequency. A closed loop op-amp circuit will have a constant closed loop gain up to the frequency at which its gain equals the open loop gain.

Figure 1-13 shows that if the op-amp's closed loop gain is 10 (20dB), its cutoff frequency will be 100KHz. If the op-amp's closed loop gain is 100 (40dB), its cutoff frequency will be 10KHz. The product of the closed loop gain and cutoff frequency is always 1MHz, the op-amp's GBP.

GBP = G_N fc. The "cut-off frequency", fc, is equal to the -3dB closed loop bandwidth. The "gain", G_N, is referred to as the "noise gain". It is equal to the inverse of the feedback factor β. It is also the gain of a non-inverting amplifier.

$$\text{Noise Gain} = Gn = \frac{Ri}{Ri+Rf} = \frac{1}{\beta}. \qquad f_C = \frac{GBP}{G_N}.$$

The closed loop bandwidth of an inverting or non-inverting op-amp is given by: fc = GBP / G_N. Note that a low gain non-inverting amplifier will have a wider closed loop bandwidth than an inverting amplifier of the same gain.

For example, if Ri = Rf, a non-inverting amplifier with a GBP of 1MHz will have a closed loop gain of 2, and a β of 0.5. Its closed loop bandwidth will be 500KHz.

An inverting amplifier with a GBP of 1MHz and a gain of 2 would have Rf = 2Ri and a β of 0.333. Its closed loop bandwidth will be 333KHz.

9

Phase Margin

Figure 1-14 below shows the typical frequency and phase response of an op-amp. A phase angle shift of 180 degrees would result in in-phase (360⁰) feedback for an inverting amplifier circuit, creating oscillation if the amplifier gain is one or greater. Figure 1-14 shows that the phase angle is -120⁰ when the amplifier gain drops to 0dB (one). The difference between 180⁰ and the phase angle where the op-amp gain drops to 0dB is called the "phase margin", 60 degrees in this case.

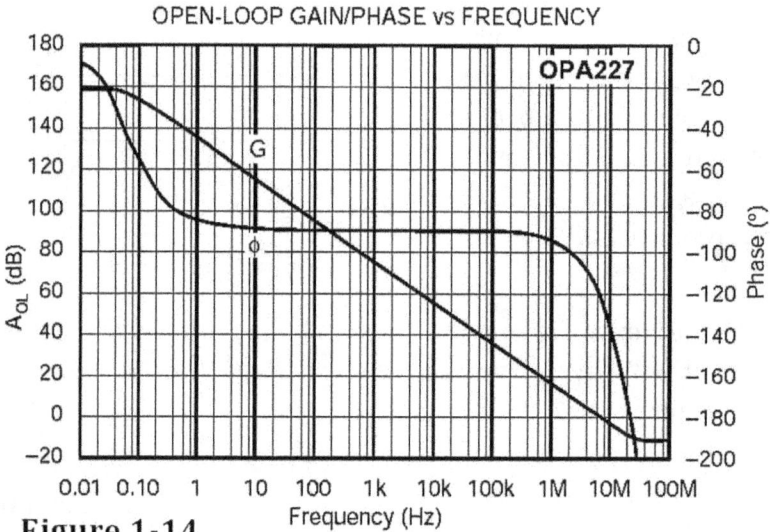

Figure 1-14

Slew Rate

An op-amp's slew rate is the maximum rate of change of its output voltage. The slew rate of a LT1013 op-amp is evaluated by simulation in figure 1-15 below. The source, V3, is a 5 volt pulse with 1 nanosecond rise and fall times. The output, vo, saturates at 4 volts and has 10 microsecond rise and fall times.

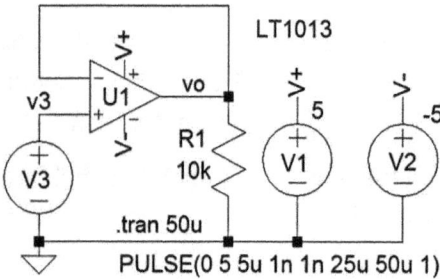

Figure 1-15

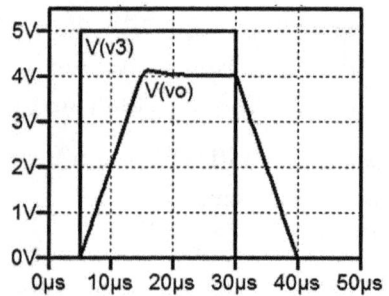

10

Slew rate of the LT1013 in the simulation example of figure 1-15:

$$\text{Slew Rate: SR} = \frac{2V}{5\mu S} = \frac{0.4V}{\mu S} = \frac{400{,}000V}{S}.$$

An op-amp's slew rate will limit its frequency response. To accurately amplify a sine wave the rate of change of the sine wave output voltage must be less than the op-amp's slew rate.

$v_o = V_p \sin(2\pi f)t.$ $\dfrac{dv_o}{dt} = V_p(2\pi f)\cos(2\pi f)t.$ $\dfrac{dv_o}{dt}$ is maximum at t = 0.

$\dfrac{dv_o}{dt} = V_p(2\pi f).$ Maximum frequency, f_m, occurs when $\dfrac{dv_o}{dt} = SR.$

$f_m = \dfrac{SR}{2\pi V_p}$, f_m is called the slew rate limited frequency.

Note that the slew rate limited frequency response is also inversely proportional to the peak amplitude of the sinusoid, V_P.

For example, for the LT1013 op-amp in figure 1-15:

If $V_p = 1 V_{peak}$, $f_m = \dfrac{SR}{2\pi V_p} = \dfrac{400{,}000}{2\pi(1)} = 63.7 KHz.$

If $V_P = 4 V_{peak}$, $f_m = \dfrac{SR}{2\pi V_p} = \dfrac{400{,}000}{2\pi(4)} = 15.9 KHz.$

The cutoff frequency of an amplifier is either the slew rate limited frequency, f_m, or the gain-bandwidth-product limited frequency, f_c, whichever is lower.

Noise

Noise voltage developed by an op-amp circuit depends on the characteristics of the op-amp and the resistances of the circuit. Resistor noise voltage is a function of the resistor value and temperature and is given by the equation below. The table gives noise voltages at 300K for a bandwidth of 1Hz.

$V_{NRT} = \sqrt{4kTB_W R}$ volts rms.

$k = 1.38 \times 10^{-23}$ Boltzmann's constant.

T = Kelvin Temperature (C + 273).

B_W = Bandwidth in Hertz.

R = Resistance in Ohms.

R Ohms	Noise Voltage RMS
10	0.41nV/√Hz
100	1.29nV√Hz
1k	4.07nV/√Hz
10k	12.9nV/√Hz
100k	40.7nV/√Hz
1Meg	129nV/√Hz

11

Amplifier Noise Model

The circuit in figure 1-16 can be used as a noise model for most inverting and non-inverting amplifiers. For an inverting amplifier, R_i would be the source resistance and R_s would be a compensating resistance. For a non-inverting amplifier, R_s would be the source resistance.

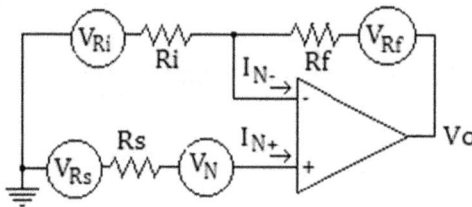

V_{Ri} is the thermal noise voltage across Ri, V_{Rs} is the thermal noise voltage across Rs, and V_{Rf} is the thermal noise voltage across Rf.

Resistors also voltage across them produced by the noise currents, I_{N+} and I_{N-}.

Figure 1-16

Noise current for an op-amp is typically specified in the op-amp's data sheet as I_N, where $I_N = I_{N+} = I_{N-}$.

The noise voltages on the non-inverting input are V_N, V_{Rs}, and $I_{N+}R_s$. V_N is the noise voltage for a particular op-amp. V_N is specified in the op-amp's data sheet. The noise voltages on the inverting input due to R_i, R_f, and I_{N-} are given below. Note: apply voltage divider rule for Ri and Rf.

$$\text{Ri: } V_{Ri}\left(\frac{R_f}{R_i+R_f}\right). \qquad \text{Rf: } V_{Rf}\left(\frac{R_i}{R_i+R_f}\right). \qquad I_{N-}:\left(\frac{R_iR_f}{R_i+R_f}\right).$$

The total noise voltage on the inputs of the op amp is given by the equation below. Note that: $I_N = I_{N+} = I_{N-}$.

$$V_{NI} = \sqrt{V_N^2+V_{Rs}^2+I_N^2R_S^2 + V_{Ri}^2\left(\frac{R_f}{R_i+R_f}\right)^2+V_{Rf}^2\left(\frac{R_i}{R_i+R_f}\right)^2+I_N^2\left(\frac{R_iR_f}{R_i+R_f}\right)^2}.$$

The total output noise voltage, V_{NO}, is equal to the total input noise voltage multiplied by the noise gain.

$$V_{NO} = V_{NI}G_N = V_{NI}\frac{R_f}{R_f+R_i}.$$

12

Differential Amplifier Noise Model

Figure 1-17 on the right shows a noise model for a differential amplifier. In this case, the input noise voltage equation is simplified because both inputs have the same resistance values.

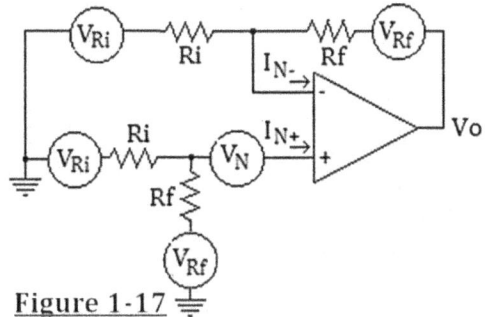

Figure 1-17

The output noise voltage is obtained by multiplying the input noise voltage by the noise gain.

$$V_{NO} = V_{NI} G_N \qquad V_{NO} = V_{NI} G_N.$$

$$V_{NI} = \sqrt{V_N^2 + 2\left\{ V_{Ri}^2 \left(\frac{R_f}{R_i + R_f} \right)^2 + V_{Rf}^2 \left(\frac{R_i}{R_i + R_f} \right)^2 + I_N^2 \left(\frac{R_i R_f}{R_i + R_f} \right)^2 \right\}}.$$

Signal-to-Noise Ratio

One measure of the quality of an electrical signal is its signal-to-noise ratio, abbreviated SNR or S/N. It is defined as the ratio of signal power to the noise power and is usually expressed in decibels. Both signal power and noise power are measured at the same circuit node and therefore across the same impedance.

$$S/N = \frac{P_{SIGNAL}}{P_{NOISE}}, \qquad S/N_{dB} = 10 log_{10} \left(\frac{P_{SIGNAL}}{P_{NOISE}} \right) = 10 log_{10} P_{SIGNAL} - 10 log_{10} P_{NOISE}.$$

S/N is often determined by measuring the signal voltage and noise voltage at the same circuit node. The S/N ratio can then be calculated directly from the voltage measurements.

$$S/N = \frac{P_{SIGNAL}}{P_{NOISE}} = \frac{V_{SIGNAL}^2 / Z}{V_{NOISE}^2 / Z} = \frac{V_{SIGNAL}^2}{V_{NOISE}^2}, \qquad \text{all voltages are in RMS units.}$$

$$S/N_{dB} = 20 log_{10} \left(\frac{V_{SIGNAL}}{V_{NOISE}} \right) = 20 log_{10} V_{SIGNAL} - 20 log_{10} V_{NOISE}.$$

13

Noise Factor

The noise factor, F, specifies the amount of S/N ratio degradation produced by devices such as a transistors, amplifiers, and receivers.

$$F = \frac{(S/N)_{INPUT}}{(S/N)_{OUTPUT}}.$$

According to IEEE Standards: "The noise factor, at a specified input frequency, is defined as the ratio of (1) the total noise power per unit bandwidth available at the output port when noise temperature of the input termination is standard (290 K) to (2) that portion of (1) engendered at the input frequency by the input termination."

$$F = \frac{\text{output noise power}}{\text{output noise power due to source}}.$$

The "output noise power due to source" is determined by multiplying the input noise power by the power gain of the circuit. A noiseless device would not add noise to the output and its noise factor would be 1. A device that adds noise to the output would have a noise factor greater than 1.

The noise factor for a cascaded system of devices with noise factors of F_1, F_2, to F_n, with gains of G_1, G_2, to G_{n-1}, is given by:

$$F = F_1 + \frac{F_2 - 1}{G_1} + \frac{F_3 - 1}{G_1 G_2} + ... + \frac{F_n - 1}{G_1 G_2 \cdots G_{n-1}}.$$

Noise Figure

The noise figure is the noise factor expressed in decibels.

$$NF_{dB} = 10 \log_{10} F.$$

Chapter 2: Op-Amp Circuit Simulation

All simulations in this book were done with *LTspice*, an easy to learn simulation program which is a free download from *Linear Technology's* website: www.linear.com. The simulations are only intended to provide examples and hints. Also refer to *LTspice's* built-in "Help". A "Getting Started Guide" and sample circuits can also be downloaded *Linear Technology's* website.

Simulation: Buffer Amplifier

Figure 2-1 shows three voltage sources, one resistor and an LT1001 op-amp. V1 and V2 are the op-amp's VCC and VEE supplies set to 10V.

V3 is connected to the non-inverting input and is set 5V DC. An operating point analysis is performed first. A DC sweep analysis is performed next. V3 is set to sweep from -10V to +10V in 0.1 volt increments.

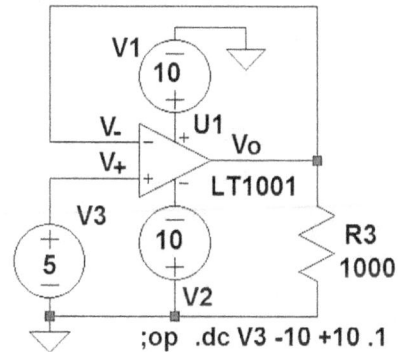

Figure 2-1

The "*Label Net*" tool was used to label the nodes V-, V+, and V.

Setup the analysis by clicking on "*Simulate*" in the main menu. Select "*Edit Simulation Cmd*". Select "*DC op pnt*". Select "*Run*".

The result shown on the right is a list of voltages and currents that represent the circuit's operating point, also called the quiescent or bias point. Note that V(V+) is equal to V3.

The value of V+ is 5V and the value of V- is 4.99999V, confirming the op-amp's input voltages are essentially equal.

```
--- Operating Point ---

V(Vo):      4.99999        voltage
V(V+):      5              voltage
V(p001):    10             voltage
V(p002):    -10            voltage
I(R3):      0.00499999     current
I(V3):      -9.02206e-012  current
I(V2):      -0.000444448   current
I(V1):      -0.00544444    current
```

Note that V- is equal to Vo because V- is directly connected to Vo. The current through R3 is 4.99999mA. Note also the very small current of 9.022 picoamps into the non-inverting op-amp input, confirming its very high input impedance.

This circuit is called a "buffer" because its input impedance is very high, output impedance is very low, and its voltage gain is equal to one. Although the voltage gain of the buffer is one, it does have a large current gain and power gain.

Setup a new analysis by clicking on "*Simulate*" in the main menu. Select "*Edit Simulation Cmd*".

Select "*DC sweep*" and enter the values as shown in the dialog box on the right.

1st Source	2nd Source	3rd Source
Name of 1st Source to Sweep:		V3
Type of Sweep:		Linear
Start Value:		-10
Stop Value:		+10
Increment:		.1

Select "<u>R</u>un". The result is shown on the right. The output voltage is equal to the input voltage except for inputs less than -8.8V or greater than +8.8V.

Region between -8.8V and +8.8 volts is called the linear region of operation.

Regions less than -8.8V and greater than +8.8 volts are called the saturation regions.

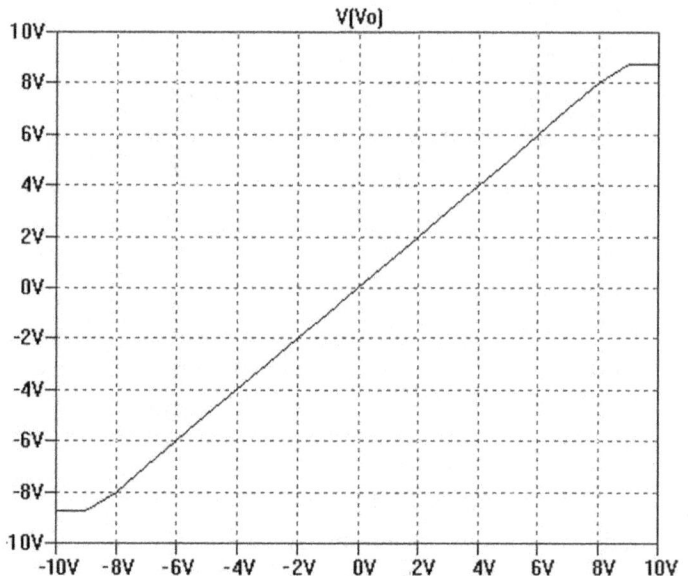
V(Vo)

The op-amp's supply voltages are also referred to as "rails". This op-amp's rails are +10V and -10V. The linear region for most op-amps does not extend to the rails as shown here. However, there are op-amps whose linear region does extend to the rails. These are called "rail to rail" op-amps. There are also op-amps whose linear region extends to the negative rail only.

Op-amps can also operate from a single power supply by biasing the inputs at one half of the supply voltage, as shown on the right. Here the non inverting input is biased by the voltage divider. The inverting input is AC coupled.

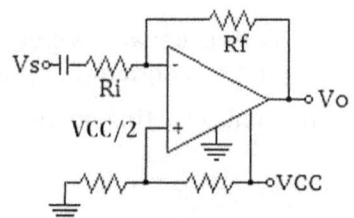

16

Simulation: Non-Inverting Amplifier

Figure 2-2 shows a non-inverting amplifier with a gain of 10. V3 is set to sweep from -1.0V to +1.0V in 0.1 volt increments.

The "*Label Net*" tool was used to label the nodes V-, V+, and Vo.

Simulate this circuit by Clicking on "*Simulate*" in the main menu. Select "*Edit Simulation Cmd*".

Select "*DC sweep*" and enter the sweep source, sweep type, and voltage values into the dialog box.

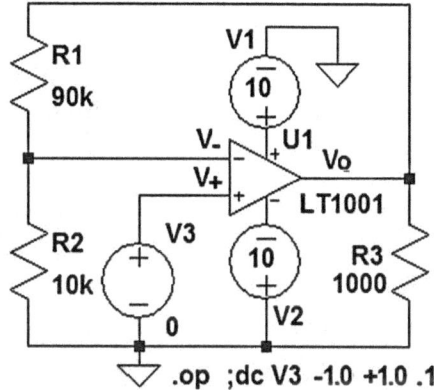

.op ;dc V3 -1.0 +1.0 .1

Figure 2-2

Select "*Run*". Select Vo in the circuit diagram to display the graph of Vo as a function of V3. Except for the gain, the graph is similar to the graph of the buffer circuit.

The output voltage is 10 times greater than the input voltage except for inputs less than about -0.8V or greater than +0.8V. The output voltage varies linearly from -8V to +8 volts. Regions less than -8.8V and greater than +8.8 volts are the saturation regions.

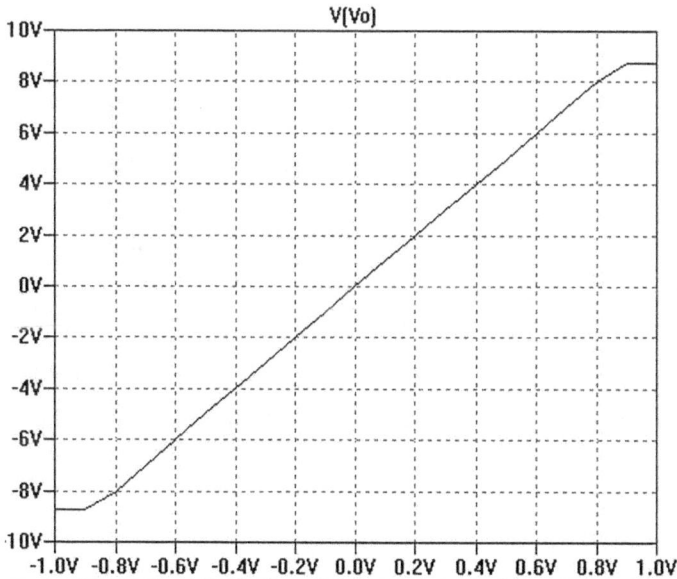

Simulation: Inverting Amplifier

Figure 2-3 on the right shows an inverting amplifier with a gain of 10. The input voltage, V3, is swept from minus 1.0V to plus 1.0V in 0.1 volt increments.

Simulate the circuit by clicking on "*Simulate*" in the main menu. Select "*Edit Simulation Cmd*". Select "*DC sweep*" and enter the sweep source, sweep type, and voltage values into the dialog box.

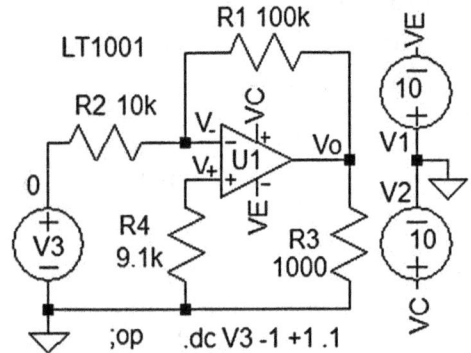

Figure 2-3

Select "Run". The result is shown below:

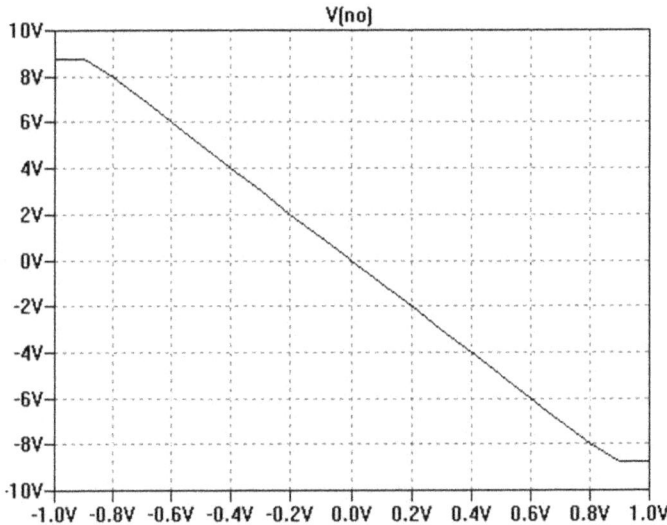

The output voltage is -10 times greater than the input voltage except for inputs less than about -0.8V or greater than +0.8V. Note that the slope of the graph is equal to the gain of the amplifier.

The output voltage varies linearly from 8V to -8 volts. This is the "linear region". The regions that are greater than 8.8V and less than -8.8 volts are the saturation regions.

Simulation: Offset Voltage and Offset Current

The *LTspice* "Opamps" library contains *Linear Technology* op-amp models, an ideal op-amp (opamp), and a generic non-ideal op-amp (UniversalOpamp2). The op-amp models in the *LTspice* library have their offset voltages and offset currents set to zero. Their bias currents are usually set to their typical data-book value.

One can refer to a particular op-amp's datasheet to determine the range of its input offset current and input offset voltage. The maximum output offset voltage error for an op-amp circuit can be calculated from this data.

When using the *LTspice* models, the offset voltages and currents can be added external to the model as shown in figure 2-4 on the right. The effect of an offset current is modeled by adding a voltage source in series with the op-amp's inverting terminal. The value of the voltage is equal to the additional voltage that would be produced by the offset current across the parallel combination of Ri and Rf.

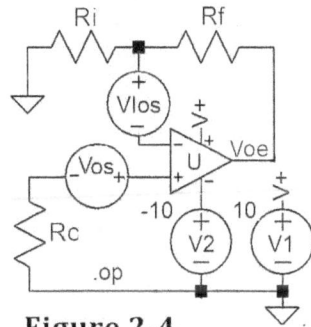

Figure 2-4

$$VIos = \frac{RiRf}{Ri+Rf} Ios.$$

A positive value of VIos will result in a positive change of the output voltage, Voe. A positive value of Vos will result in a positive change of the output voltage, Voe.

It's easy enough to calculate the effect of an offset voltage or offset current in a particular circuit. The LTspice circuit in figure 2=4 above can be used to check calculated results. Refer to the example below.

Example Results: LT1013

A *Linear Technology* LT1013 was used in the circuit of figure 2-4. Simulations were run for various resistor values, input offset voltages, and input offset currents. The LT1013's offset voltage was determined by setting Ri, Rc, VIos and Vos to zero.

For this simulation Ri and Rc were actually set to 1μΩ so that they would still appear on the schematic and in the operating point analysis.

LTspice analysis results are given in the table below.

Rf=2k except in last row Rf=2M	Voe	I(Ri)	I(Rc)	I(Rf)
1 Ri=Rc=1μ, Vios=0, Vos=0	-25pV	12nA	12nA	-.13fA
2 Ri=2k, Rc=1K, Vios=0, Vos=0	12pV	6nA	12nA	-6nA
3 Ri=2k, Rc=1K, Vios=0, Vos=50m	100mV	25μA*	12nA	-25μA*
4 Ri=2k, Rc=1K, Vios=8μ, Vos=0	16μV	10nA	12nA	-2nA
5 Ri=2M, Rc=1M, Vios=8m, Vos=0	16mV	10nA	12nA	-2nA
*I(Ri) = 25006nA, *I(Rf) = -24994nA, I(Ri) - I(Rf) = 12nA				

Row 1: The LT1013's input offset voltage is zero because the output voltage, Voe, is essentially zero (-25 X 10^{-12} volts). There can be no output voltage error due to bias currents because both inputs are connected to ground (through 1μΩ resistors).

Row 2: The LT1013's non-inverting input's bias current is 12nA as indicated by I(Rc). Ri and Rf each have a current of 6nA. Both of these currents flow into the LT1013's inverting input.   So both inputs have the same bias current indicating that the offset current is zero. Since the value of Rc is equal to the value of the parallel combination of Ri and Rf, the output offset error voltage, Voe, is essentially zero.

Row 3: An input offset voltage of 50mV produces an output error voltage of 100mV. Note that the bias currents are still 12nA. The theoretical output error voltage is calculated below.

$$Voe = Vos\left(\frac{Ri+Rf}{Ri}\right) = 50mV\left(\frac{2k+2k}{2k}\right) = 100mV.$$

Row 4: An input offset current of 8nA (represented by VIos = 8μV) produces an output error voltage of 16μV.

$$VIos = \frac{RiRf}{Ri+Rf}Ios = (1k)(8nA) = 8μV.$$

Row 5: If all the resistor values are increased by a factor of 1000, the output offset voltage also increases by a factor of 1000. The theoretical output error voltage is calculated below.

$$Voe = Rc\left(Ib_- - Ib_+\right)\left(\frac{Ri+Rf}{Ri}\right) = (1 \times 10^6)(8 \times 10^{-9})2 = 16mV.$$

Simulation: Op-Amp Noise

Spice op-amp models will usually have typical data sheet noise values.

The LT1013 data sheet specifies the op-amp's equivalent input noise voltage, V_{NI}, at 10Hz as 24nV/√Hz. Its equivalent input noise current, I_N, at 10 Hz is 0.07pA/√Hz. These values are per unity square root bandwidth.

.noise V(Vo) V1 dec 10 10 1meg

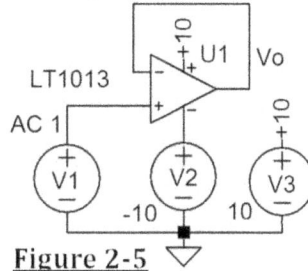

Figure 2-5

Output:	V(Vo)
Input:	V1
Type of Sweep:	Decade ∨
points per decade:	10
Start Frequency:	10
Stop Frequency:	1meg

An op-amp model's equivalent input noise voltage can be determined by performing a noise simulation on a voltage follower (buffer) circuit.

The circuit in figure 2-5 is simulated. Select the "Edit Simulation Cmd". Select "Noise". The dialog box shown on the left appears.

The noise simulation settings for this example are shown above and the simulation result is shown below.

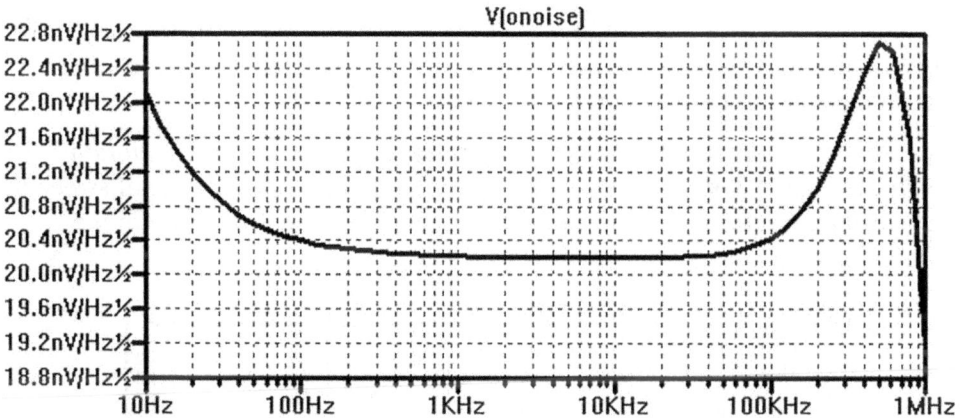

The noise simulation result above shows 22nV/√Hz at 10Hz which is close to the data sheet value of 24nV/√Hz.

Simulation: Frequency Response and Noise

Figure 2-6 on the right shows an inverting amplifier with a gain of 100. Since the gain-bandwidth-product of the LT1013 is 1MHz, its cutoff frequency is about 10KHz.

First, the frequency response of the amplifier will be simulated. Select "*Edit Simulation Command*" and "*AC Analysis*". Set type of sweep to decade, points per decade to 10, start frequency to 10, and stop frequency to 1meg. The result is shown below. The dashed line is the phase response.

;noise V(Vo) V1 dec 10 10 1meg
.ac dec 10 10 1meg

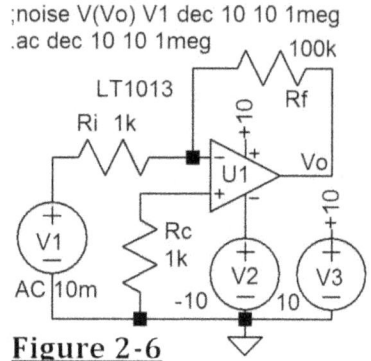

Figure 2-6

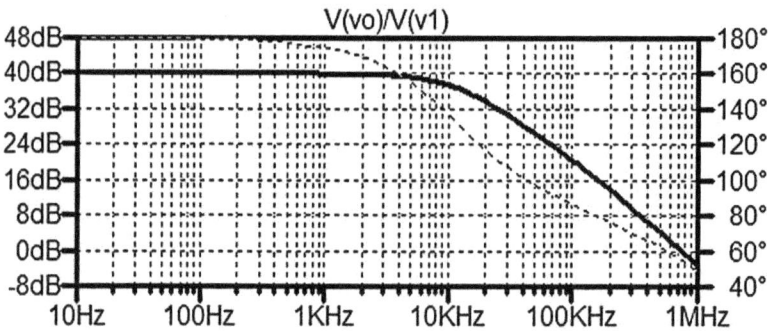

V(vo)/V(v1)

AC analysis frequency response plot verifies the expected cutoff frequency of 10KHz. It also shows that the unity gain frequency is 100KHz and that the circuit's phase margin is about 55 degrees.

Next, the noise of the amplifier will be simulated. Select "*Edit Simulation Command*" and "*Noise*". Set the output to V(Vo), input to V1, type of sweep to decade, start frequency to 10, and stop frequency to 1meg. A graph of the simulation results is shown below.

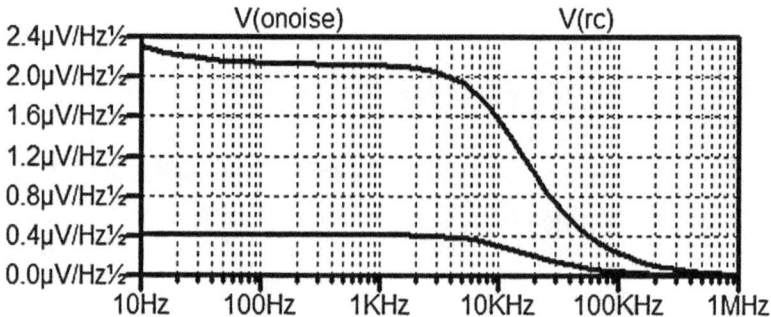

V(onoise)   V(rc)

The plot above shows the total noise and the noise for the resistor, Rc.

Calculated noise

Resistor thermal noise: 1k = 4.07nV/√Hz,   100k = 40.7nV/√Hz.
Resistor current noise: 1k =.07nV/√Hz. Current noise is negligible.
Total noise, V_{NI}:

$$V_{NI} = \sqrt{(22nV)^2 + (4.07nV)^2 + (4.07nV)^2(.99)^2 + (40.7nV)^2(.0099)^2}.$$

$$V_{NI} = 22.74nV, \quad V_{No} = 22.74nV(101) = 2.3\mu V.$$

Simulation: Transient Response

Transient analysis is used to display circuit voltages and currents in the time-domain. The voltage source V1 in figure 2-7 on the right is set by right clicking on it. V1 is set to produce a 400mV, 1KHz sine wave as shown below.

SINE(0 400m 1k 0 0 0 2.5)   .tran 2.5m

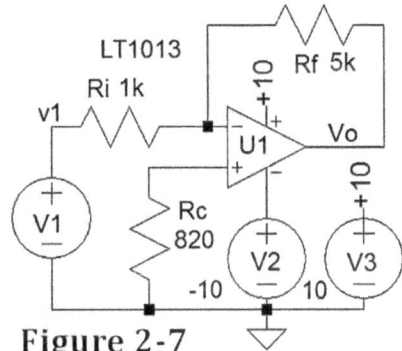

Figure 2-7

Functions
- ○ (none)
- ○ PULSE(V1 V2 Tdelay Trise Tfall Ton Period Ncycles)
- ⦿ SINE(Voffset Vamp Freq Td Theta Phi Ncycles)
- ○ EXP(V1 V2 Td1 Tau1 Td2 Tau2)
- ○ SFFM(Voff Vamp Fcar MDI Fsig)
- ○ PWL(t1 v1 t2 v2...)
- ○ PWL FILE: B

DC offset[V]:	0
Amplitude[V]:	400m
Freq[Hz]:	1k
Tdelay[s]:	0
Theta[1/s]:	0
Phi[deg]:	0
Ncycles:	2.5

Two and a half cycles of the input and output wave forms are show in the result of the simulation below.

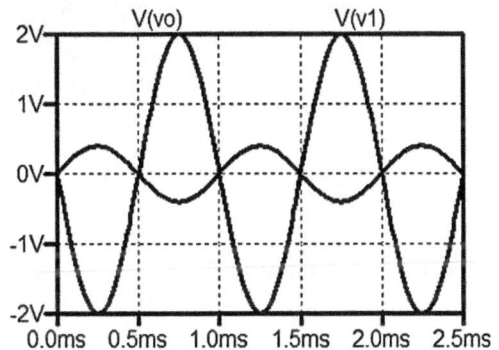

This is a gain 5 inverting amplifier. Vo is 5 times V1 and 180° out of phase.

23

Chapter 3: Filters

Low-Pass Filters

Low-pass filters are used to pass frequencies below a parameter called the "cutoff frequency", and to reject frequencies above the cutoff frequency. The cutoff frequency is defined as the frequency where the power output of the filter is one half of the power input. This occurs when the magnitude of the output voltage is equal to the magnitude of the input voltage divided by the square root of two.

The magnitude of the transfer function and the cutoff frequency of the single pole low-pass filter shown in Figure 3-1 are given by:

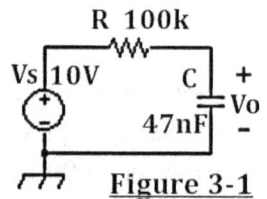

R 100k

Vs 10V   C   +
47nF   Vo   -

Figure 3-1

$$\left|\frac{V_o}{V_s}\right| = \frac{1}{\sqrt{1+(\omega/\omega_c)^2}}. \qquad \omega_c = \frac{1}{RC} = 212.8 r/s.$$

The circuit in figure 3-2 on the right shows an op-amp single pole low-pass filter. It has exactly the same frequency response as the RC filter in figure 3-1, except that it has an 180^0 phase inversion. However the op-amp version has very low output impedance and it can have a voltage gain. But it does need a power supply. The transfer function of the op-amp filter is derived below using the node voltage method in the phasor domain.

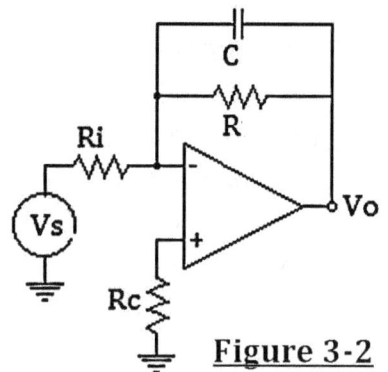

C

R

Ri

Vs

Rc

Vo

Figure 3-2

$$\frac{0-Vs}{Ri}+\frac{0-Vo}{R}+\frac{(0-Vo)\omega C}{-j}=0 \quad \Rightarrow \quad VsR+VoRi+jVo\omega RCRi=0.$$

$$Vo(Ri+j\omega RCRi)=-VsR \quad \Rightarrow \quad \left|\frac{Vo}{Vs}\right|=\frac{-R}{Ri+j\omega RCRi}=\left(\frac{-R}{Ri}\right)\left(\frac{1}{1+j\omega RC}\right).$$

The magnitude of the transfer function and the cutoff frequency of the single pole low-pass filter shown in Figure 3-2 are given by:

$$\left|\frac{V_o}{V_s}\right|=\frac{-A_{INV}}{\sqrt{1+(\omega/\omega_c)^2}}, \qquad \omega_c=\frac{1}{RC}, \qquad \theta=180-\arctan\frac{\omega}{\omega_c}.$$

If R = 100k, Ri = 10K, C = 49nF: $A_{INV}=10$, $\omega_c=212.8 r/s$.

The frequency response of the op-amp low-pass filter is shown below.

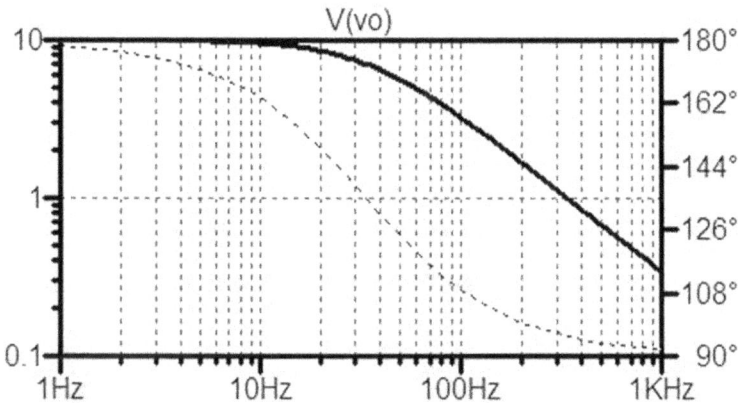

Vs is 1V. Vo is 7.07 volts (-3 dB) at the cutoff frequency of 34 hertz (213r/s). Vo decreases by a factor of 10 for each decade increase of frequency beyond the cutoff frequency (-20 dB/decade).

Second Order Low-Pass Filter

Figure 3-3 shows an *LTspice* circuit diagram of a 2-pole low-pass filter. It uses two RC filter sections, R1C1 and R2C2. The cutoff frequency of each filter section is 34Hz, the same as the single pole filter in figure 3-2.

Right click on V1 to set its AC amplitude to 1V. Select AC Analysis.

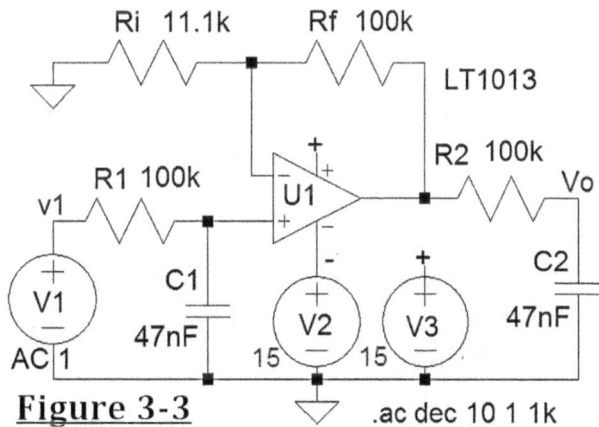

Figure 3-3

Set sweep type to decade, 10 points/decade, start frequency to 1Hz, and stop frequency to 1KHz. The result is shown on the right.

The filter's cutoff frequency is about 22Hz. Vo decreases by a factor of 100/decade above the cutoff frequency (-40 dB/dec).

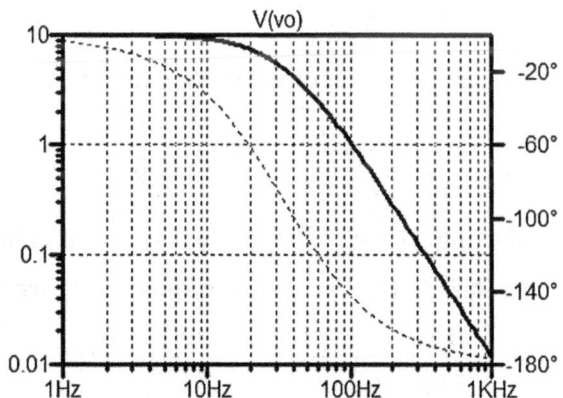

Values for the filter's frequency and phase response are calculated below.

Let R1 = R2, C1 = C2, and $\omega_X = \dfrac{1}{R1C1} = \dfrac{1}{R2C2}$.

$$\left|\frac{Vo}{V1}\right| = \frac{A_{NI}}{\sqrt{1+\left(\omega/\omega_X\right)^2}} \frac{1}{\sqrt{1+\left(\omega/\omega_X\right)^2}} = \frac{A_{NI}}{\left(1+\left(\omega/\omega_X\right)^2\right)}, \quad \theta = -2\left(\arctan\frac{\omega}{\omega_C}\right).$$

If Rf = 100k, Ri = 11.1K, R1=R2=100k, C1 = C2=49nF: A_{NI}=10, ω_X=212.8r/s.

The cutoff frequency occurs when $\left|Vo\right| = 1/\sqrt{2}$ of its maximum value.

$$\frac{1}{\left(1+\left(\omega_C/\omega_X\right)^2\right)} = \frac{1}{\sqrt{2}} \Rightarrow \left(\frac{\omega_C}{\omega_X}\right)^2 = \sqrt{2}-1 \Rightarrow \omega_C = .64\,\omega_X = \frac{.64}{R1C1} = 136.2r/s.$$

This filter has a cutoff frequency of 21.7Hz, which is lower by a factor of 0.64 than the cutoff frequencies of the individual sections.

Second Order Low-Pass Butterworth Filter

Butterworth filters have a flatter pass-band response and a better attenuation beyond their pass-band. Figure 3-4 shows an *LTspice* circuit diagram for a 2-pole, low-pass, Butterworth filter.

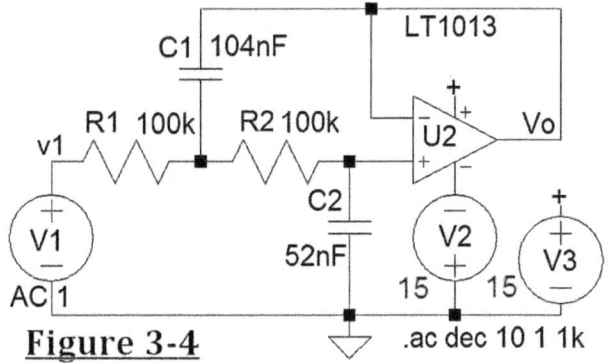

Figure 3-4

The response of order "n" low-pass Butterworth filter is calculated below.

$$\left|\frac{Vo}{v1}\right| = \frac{1}{\sqrt{1+\left(\omega/\omega_C\right)^{2n}}} \qquad \omega_C \text{ is the filter's cutoff frequency.}$$

The calculations for a 2-pole filter in figure 3-5 are presented below.

$$\left|\frac{Vo}{v1}\right| = \frac{1}{\sqrt{1+\left(\omega/\omega_C\right)^4}} \qquad \omega_C = \frac{1}{\sqrt{2}RC2} \quad \text{if } R1 = R2 = R \text{ and } C1 = 2C2.$$

Selecting R1 = R2 = 100k, the values of C1 and C2 are calculated for a cutoff frequency of 136.2r/s (21.7Hz, same as figure 3-4) are calculated below.

$$C2 = \frac{1}{\sqrt{2}R\omega_C} = \frac{1}{\sqrt{2}\left(1\times10^5\right)(136.2)} = 52nF. \qquad C1 = 2C2 = 104nF.$$

Below: The frequency response of the Butterworth filter in figure 3-4 is compared to the frequency response of the cascaded filter in figure 3-3.

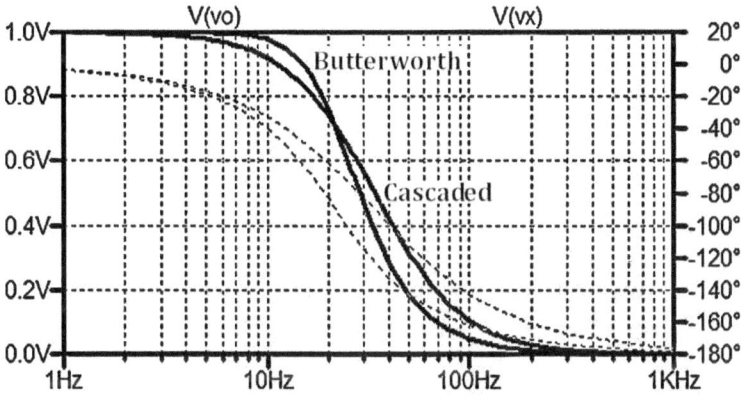

The Butterworth filter has a much flatter response in the pass band and greater attenuation beyond the cutoff frequency.

High-Pass Filters

High-pass filters are used to pass frequencies above a parameter called the "cutoff frequency", and to reject frequencies below the cutoff frequency. The cutoff frequency is defined as the frequency where the power output of the filter is half of the power input. This occurs when the magnitude of the output voltage is equal to the magnitude of the input voltage divided by the square root of two.

The transfer function and cutoff frequency of the circuit in figure 3-5 on the right are given by:

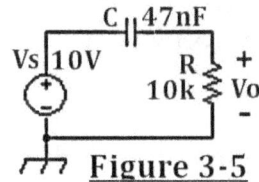

Figure 3-5

$$\left|\frac{Vo}{Vs}\right| = \frac{1}{\sqrt{1+(\omega_c/\omega)^2}}.$$

$$\omega_c = \frac{1}{RC} = 2128 r/s = 338.7\,Hz.$$

$$\theta = -\arctan(\omega_c/\omega).$$

The circuit in figure 3-6 on the right shows an op-amp single pole high-pass filter. It has exactly the same frequency response as the RC filter in figure 3-5 except for a 180 phase inversion.

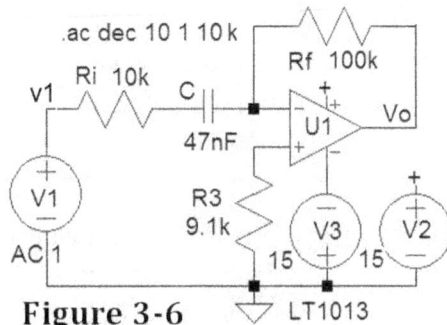

Figure 3-6

The voltage transfer function of the op-amp filter is derived below using the node voltage method in the phasor domain.

$$\frac{0-V1}{R-j\left(\frac{1}{\omega C}\right)}+\frac{0-Vo}{Rf}=0 \quad \Rightarrow \quad V1\,Rf=-Vo\left(R-j\left(\frac{1}{\omega C}\right)\right).$$

$$\left|\frac{Vo}{V1}\right|=\frac{-Rf}{\left(R-j\left(\frac{1}{\omega C}\right)\right)}=\frac{-Rf\!\!\Big/R}{\left(1-j\left(\frac{1}{\omega RC}\right)\right)}=\frac{-A_{INV}}{\left(1-j\left(\frac{\omega_C}{\omega}\right)\right)}.$$

The magnitude of the transfer function and the cutoff frequency of the single pole high-pass filter shown in Figure 3-6 are given by:

$$\left|\frac{Vo}{V1}\right|=\frac{-A_{INV}}{\sqrt{1+\left(\omega_C/\omega\right)^2}}, \quad \omega_C=\frac{1}{RC}, \quad \theta=180+\arctan\frac{\omega_C}{\omega}.$$

If $Rf = 100k$, $R = 10K$, $C = 49nF$: $A_{INV}=10$, $\omega_C=2128r/s$.

The frequency response of the filter is shown on the right.

Vo is 7.07 volts (-3 dB) at the cutoff frequency of 338 hertz. Vo decreases by a factor of 10 for each decade decrease of frequency beyond the cutoff frequency (-20 dB/decade).

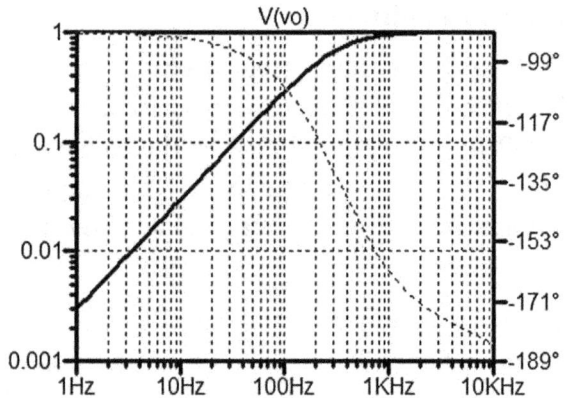

Second Order High-Pass Filter

Higher order high-pass filters can be made by cascading single pole filters. In this case the cascaded filter's cutoff frequency higher than the cutoff frequency of the individual sections.

An *LTspice* schematic diagram of a cascaded 2-pole high-pass filter is shown in figure 3-7 on the right.

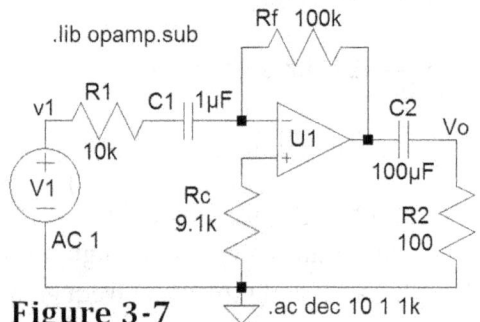

Figure 3-7

Both sections of this filter have the same cutoff frequency except that the input filter has an impedance of 10kΩ and the output filter has an impedance of 100Ω. The filter's response is calculated below.

$$\left|\frac{Vx}{V1}\right|=\frac{A_{INV}}{\sqrt{1+\left(\omega_{C1}/\omega\right)^2}}, \quad \omega_{C1}=\frac{1}{R1C1}, \quad \theta_1=180+\arctan\frac{\omega_{C1}}{\omega}.$$

$$\left|\frac{Vo}{Vx}\right|=\frac{1}{\sqrt{1+\left(\omega_{C2}/\omega\right)^2}}, \quad \omega_{C2}=\frac{1}{R2C2}, \quad \theta_2=\arctan\frac{\omega_{C2}}{\omega}.$$

If R1C1=R2C2, then $\omega_{C1}=\omega_{C2}=\omega_c$. The filter's reponse can be calculated as:

$$\left|\frac{Vo}{V1}\right|=\left|\frac{Vx}{V1}\right|\left|\frac{Vo}{Vx}\right|=\frac{A_{INV}}{\sqrt{1+\left(\omega_x/\omega\right)^2}}\frac{1}{\sqrt{1+\left(\omega_x/\omega\right)^2}}=\frac{A_{INV}}{1+\left(\omega_x/\omega\right)^2}.$$

$$\theta=\theta_1+\theta_2=180^0+\arctan\frac{\omega_x}{\omega}+\arctan\frac{\omega_x}{\omega}=180^0+2\arctan\frac{\omega_x}{\omega}.$$

If Rf = 100k, R1 = 10K, C1 = 1uF, R2 = 100, C1 = 100uF:

$A_{INV}=10$, $\omega_x=100r/s = 15.9Hz$.

The cutoff frequency, ω_c, occurs when $|Vo|=1/\sqrt{2}$ of its maximum value.

$$\frac{1}{1+\left(\omega_x/\omega_c\right)^2}=\frac{1}{\sqrt{2}} \Rightarrow \left(\frac{\omega_x}{\omega_c}\right)^2=.64 \Rightarrow \omega_c=\frac{\omega_x}{.64}=\frac{100}{.64}=156r/s=24.9Hz.$$

$$\theta_c=180^0+2\arctan\frac{\omega_x}{\omega_c}=180^0+2\arctan\frac{100}{156}=-180^0+65.3^0=-114.7^0.$$

The simulation result for the circuit in figure 3-7 is shown on the right.

Cutoff frequency is about 25Hz. The Phase angle at the cutoff frequency is about -115⁰.

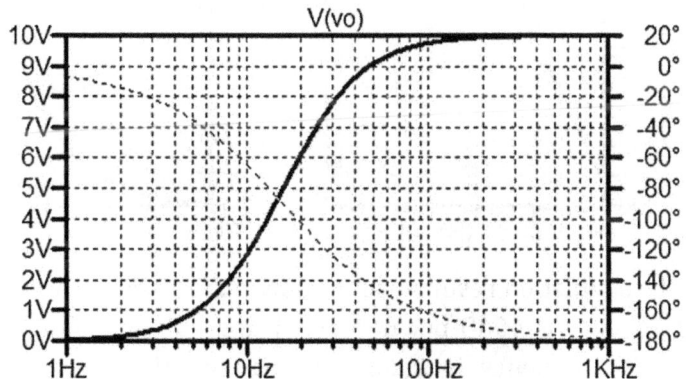
V(vo)

Second Order High-Pass Butterworth Filter

An *LTspice* schematic diagram high-pass Butterworth filter is shown in figure 3-8. Given that R2 = 2R1 and C1 = C2, the cutoff frequency, f_c, is given by:

$$f_c = \frac{1}{2\pi\sqrt{2}CR1}.$$

Resistor values should be between 10k and 100k ohms for best performance.

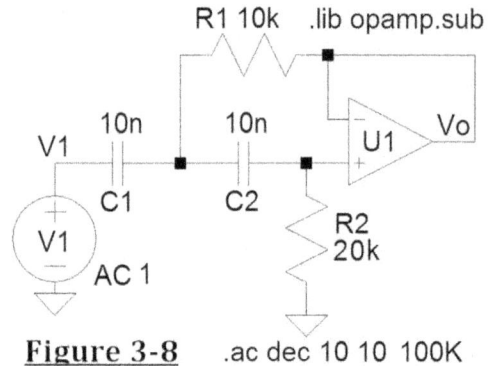

Figure 3-8   .ac dec 10 10 100K

$$\left|\frac{Vo}{v1}\right| = \frac{1}{\sqrt{1+(\omega_c/\omega)^{2n}}} \qquad \omega_c \text{ is the filter's cutoff frequency.}$$

The calculations for a 2-pole filter in figure 3-8 are presented below.

$$\left|\frac{Vo}{v1}\right| = \frac{1}{\sqrt{1+(\omega_c/\omega)^{4}}} \qquad \omega_c = \frac{1}{\sqrt{2}CR1} \quad \text{if } R2 = 2R1 \text{ and } C1 = C2 = C.$$

Given R1 = 10k, R2 = 20k, C1 = C2 = 10nF, the cutoff frequency is calculated below.

$$\omega_c = \frac{1}{\sqrt{2}CR1} = \frac{1}{\sqrt{2}(10nF)(10k)} = 7071 r/s = 1125Hz.$$

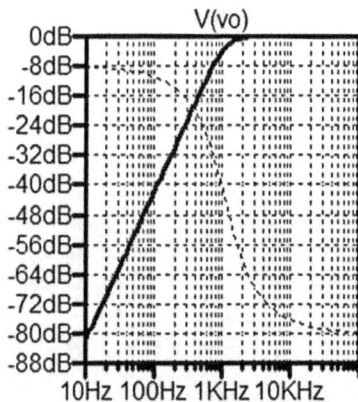

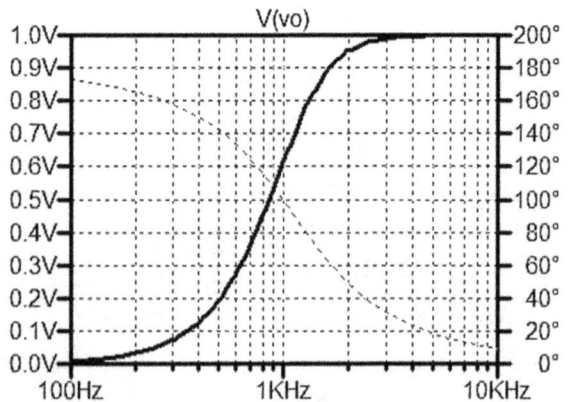

The simulation results above show that the filters attenuation below the cutoff frequency is 40dB per decade. The graph above on the right shows that the cutoff frequency is about 1100Hz and that the phase angle at the cutoff frequency is about 90 degrees.

Band-Pass Filter

The operational amplifier band-pass filter circuit on the right uses a high-pass and a low-pass filter section to obtain a band-pass response.

The RC network, R2 and C2, in its negative feedback path provides the low-pass portion of the band-pass response. Series connected C1 and R1 provide the high- pass portion.

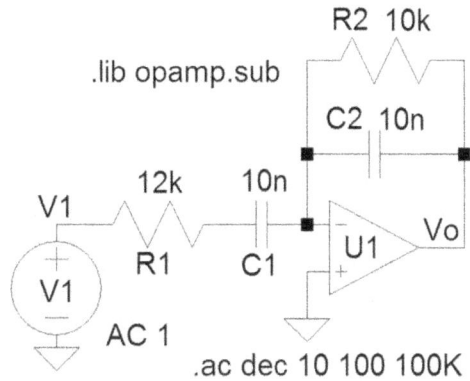

R2 10k

.lib opamp.sub

C2 10n

12k   10n

V1

R1   C1   U1   Vo

V1

AC 1

.ac dec 10 100 100K

Figure 3-9

The node voltage method is used to find the circuit's transfer function:

$$\frac{0-\text{Vin}}{R1-j\dfrac{1}{\omega C1}}+\frac{0-\text{Vout}}{R2}+\frac{0-\text{Vout}}{-j\dfrac{1}{\omega C2}}=0. \quad \text{Vout}\left(1+\frac{1}{-j\dfrac{1}{\omega R2C2}}\right)=\text{Vin}\left(\frac{-\dfrac{R2}{R1}}{1-j\dfrac{1}{\omega R1C1}}\right).$$

$$\frac{\text{Vout}}{\text{Vin}}=\frac{-\left(\dfrac{1}{R1C2}\right)j\omega}{\left(j\omega+\dfrac{1}{R2C2}\right)\left(j\omega+\dfrac{1}{R1C1}\right)}=\frac{\left(-j\omega\dfrac{1}{R1C2}\right)}{-\omega^2+j\omega\left(\dfrac{1}{R1C1}+\dfrac{1}{R2C2}\right)+\left(\dfrac{1}{R1C1}\right)\left(\dfrac{1}{R2C2}\right)}.$$

$$\frac{\text{Vout}}{\text{Vin}}=\frac{-j\omega K\omega_2}{\left(\omega_0^2-\omega^2\right)+j\omega\left(\omega_1+\omega_2\right)} \quad K=\frac{R2}{R1} \quad \omega_1=\frac{1}{C1R1} \quad \omega_2=\frac{1}{C2R2} \quad \omega_0^2=\omega_1\omega_2$$

Gain when $\omega=\omega_0$: $A_V=\dfrac{-K\omega_2}{\left(\omega_1+\omega_2\right)}$, Bandwidth: $B_W=\omega_1+\omega_2$, $\omega_0=\sqrt{\omega_1\omega_2}$

Given the part values in figure 3-9:

$$\omega_1=\frac{1}{R1C1}=\frac{1}{(12k)(10n)}=8333\,\text{r/s}=1326\text{Hz}.$$

$$\omega_2=\frac{1}{R2C2}=\frac{1}{(10k)(10n)}=10000\,\text{r/s}=1591\text{Hz}.$$

$\omega_0=\sqrt{\omega_1\omega_2}=9128\,\text{r/s}=1453\text{Hz}, \quad B_W=\omega_1+\omega_2=1326\text{Hz}+1591\text{Hz}=2917.$

The actual cutoff frequencies are the frequencies where the magnitude of the filter's transfer function is equal to -3 dB of its maximum value. The theoretical values of these frequencies are most easily found by simulation.

31

Simulation of Band-Pass Filter

This simulation uses the generic "*opamp*" in the opamp library. The directive, "*.lib opamp.sub*" must be added. Click on "*op*" on the right side of the main menu bar and type in the directive. It will be displayed on the schematic as shown below.

This simulation analysis was set to "*AC Analysis*", decade sweep, 50 points per decade, start frequency = 100, and stop frequency = 100,000. Right click on the voltage source to open its dialog box. Set AC amplitude to 1. Simulation result is shown below.

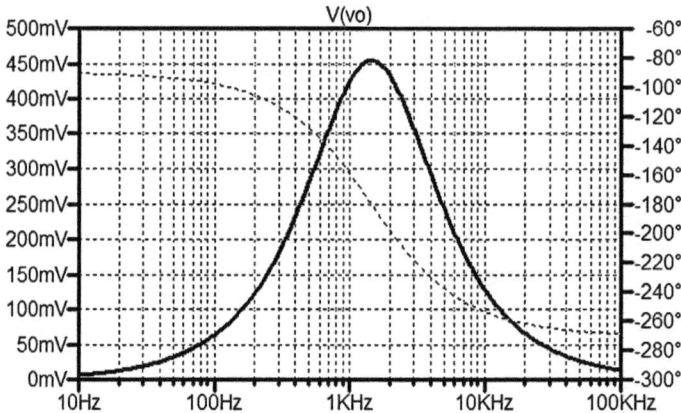

Cursors are obtained by right clicking on V(vo) at the top of the graph. Select "*Attached Cursor 1st and 2nd*". Move the mouse over the plot to read the x-y coordinates of the cursor.

The result is shown on the right. Cursor 1 was set on ω_0, where the amplitude is maximum. Cursor 2 was set to the low cutoff frequency where the amplitude is 0.707 of maximum. Cursor 2 was moved to the high cutoff frequency where the amplitude is 0.707 of maximum.

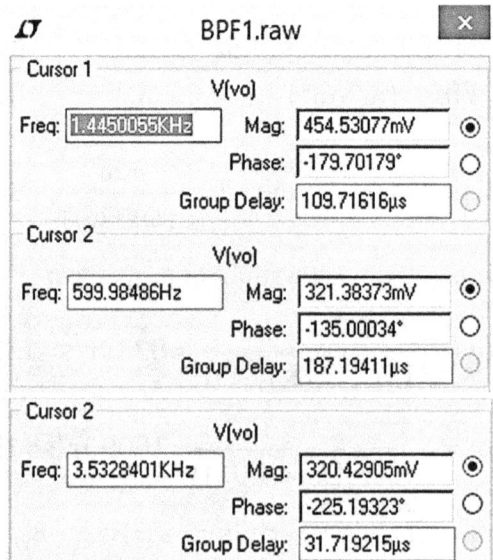

BPF1.raw

Cursor 1

V(vo)

Freq: 1.4450055KHz   Mag: 454.53077mV
Phase: -179.70179°
Group Delay: 109.71616µs

Cursor 2

V(vo)

Freq: 599.98486Hz   Mag: 321.38373mV
Phase: -135.00034°
Group Delay: 187.19411µs

Cursor 2

V(vo)

Freq: 3.5328401KHz   Mag: 320.42905mV
Phase: -225.19323°
Group Delay: 31.719215µs

The results show that the resonant frequency is 1445Hz and the bandwidth is 2933Hz. The calculated values are 1453Hz for the resonant frequency and 2917Hz for bandwidth.

Infinite-Gain Multiple-Feedback Band-Pass Filter

Figure 3-10 shows an *LTspice* schematic diagram of an infinite-gain multiple-feedback (IGMF) band-pass filter. This filter is capable of a narrow band tuned response and may have a quality factor "Q" of up to about 20. It is specified by a center frequency ω_0, and a bandwidth, B_W.

For best results, limit the bandwidth and Q to about 10.

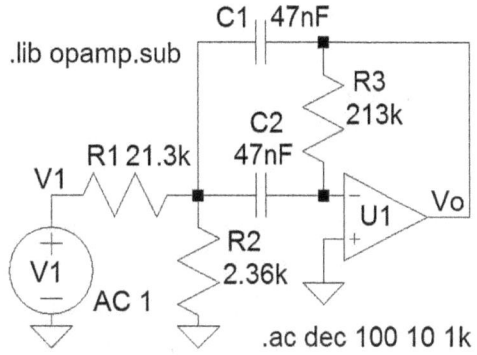

Figure 3-10

$$A_0 = -\left(\frac{R_3}{2R_1}\right). \qquad \omega_0 = \frac{1}{C\sqrt{(R_1\|R_2)R_3}}. \qquad B_W = \frac{2}{R_3C}. \qquad Q = \frac{\omega_0 CR_3}{2}.$$

Select $\omega_0 = 1000 r/s\,(159Hz)$, $B_W = 200 r/s\,(31.8Hz)$, $A_0 = 5$. Pick C: 47nF.

$$R_3 = \frac{2}{B_W C} = \frac{2}{200(47\times10^{-9})} = 213k.$$

$$\sqrt{(R_1\|R_2)} = \frac{1}{\omega_0 C\sqrt{R_3}} = \frac{1}{1k(47n)\sqrt{213k}} = 46. \qquad R_1 = \frac{R_3}{2A_0} = \frac{213k}{10} = 21.3k.$$

$$\frac{R_1 R_2}{R_1 + R_2} = 46^2 = 2125 \quad \Rightarrow \quad (21.3k)R_2 = 2125(21.3k) + 2125R_2.$$

$$19.175R_2 = 45263. \qquad R_2 = 2.36k.$$

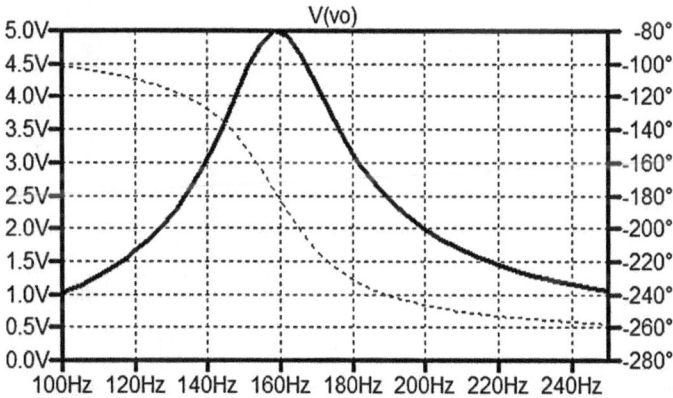

The simulation results above show that the filters response parameters, ω_0, B_W, and A_0, are very close to the calculated values.

Notch Filter

Figure 3-11 shows an *LTspice* diagram of a notch filter circuit using a single op amp. This notch filter is capable of notch frequency attenuation of about 50dB.

One percent tolerance or better, resistor and capacitor values must be used. This filter is commonly used to suppress 50Hz or 60Hz line interference. Values are calculated for a 60Hz notch below.

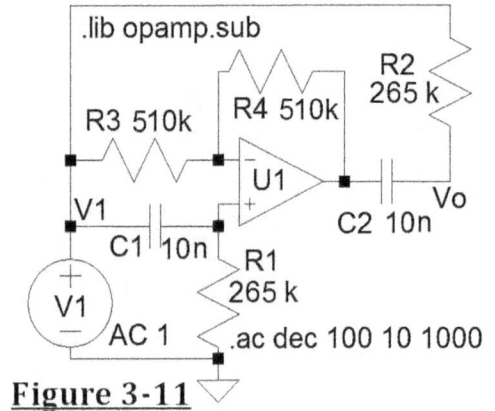

Figure 3-11

$$R1=R2=R, \quad C1=C2=C, \quad \omega_0 = \frac{1}{RC}. \quad \omega_0 = 377\,r/s. \quad \text{If } C=10nF, \quad R=\frac{1}{\omega_0 C}=267k.$$

AC analysis was set to decade, 100 points per decade, start frequency 10Hz, stop frequency 1kHz. Right click on V1 to set its AC amplitude to 1 volt.

The LTspice simulation result for this 60Hz notch filter is shown below.

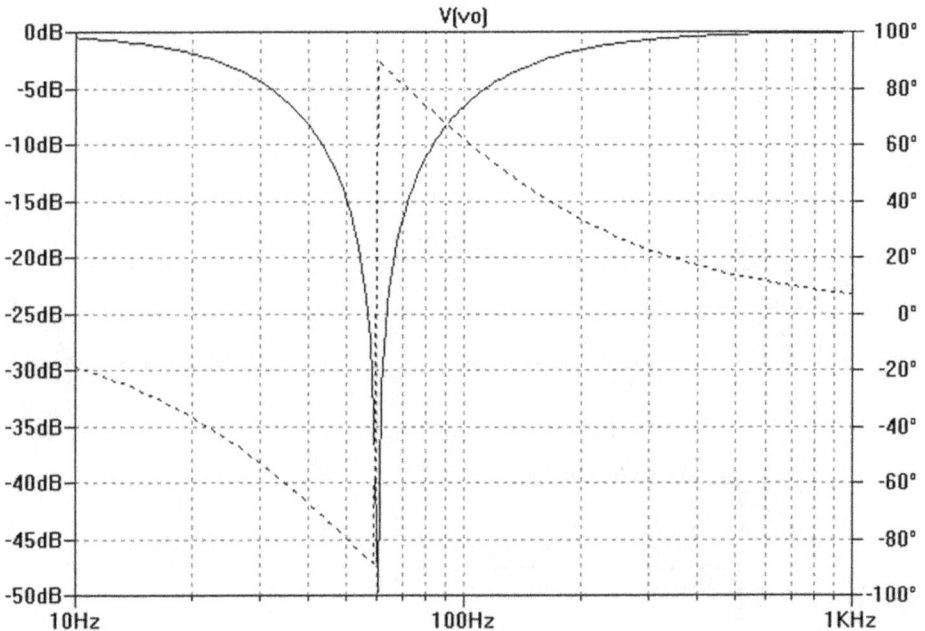

34

Twin-T Notch Filter

The twin-T filter in figure 3-12 is commonly used between two op-amps; the first op-amp provides a very low impedance source for the filter's input and the second op-amp provides a very high impedance load for the filter's output.

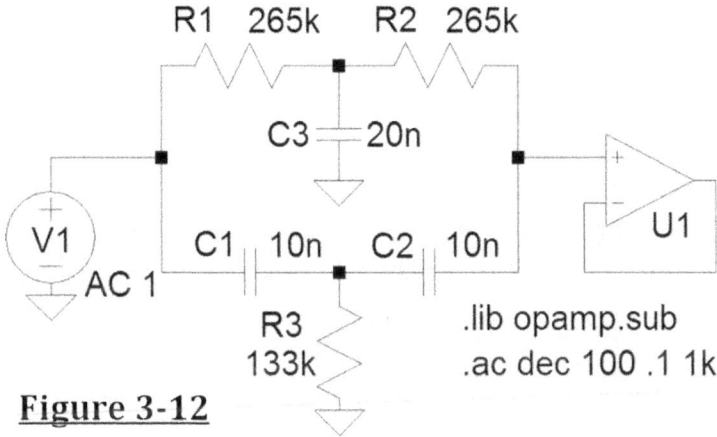

Figure 3-12

A notch of about -60dB can be attained using precision resistors and capacitors. However, this filter requires 6 precisely matched components. Also, its -3dB bandwidth is rather large, 226Hz for the one below.

AC analysis was set to decade, 100 points per decade, start frequency 0.1Hz, stop frequency 1kHz. Right click on V1 to set its AC amplitude to 1 volt.

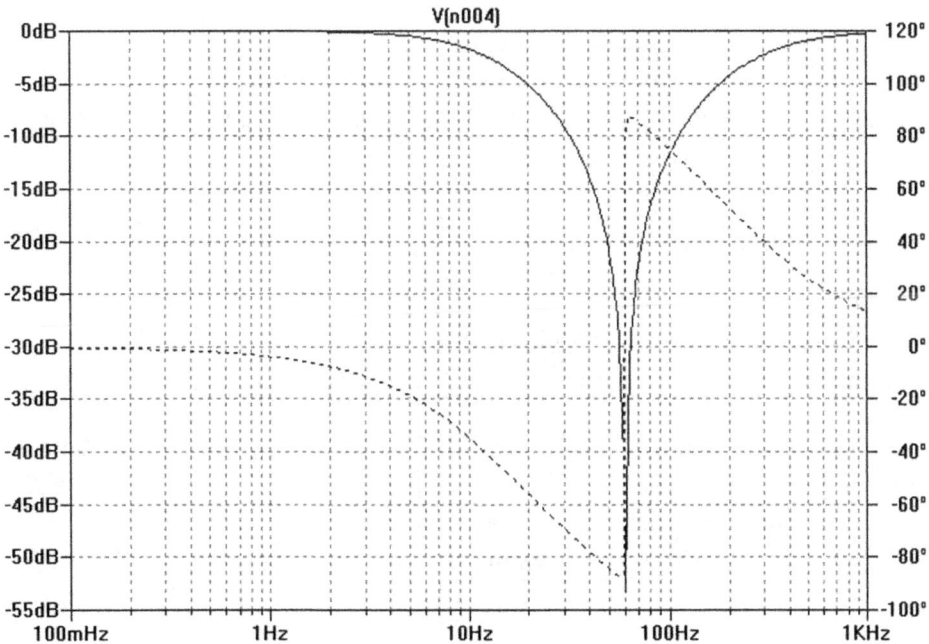

This filter's -3dB frequencies are 14Hz and 240Hz. The equation for the filter's notch frequency is given below.

$$\text{If } R1=R2=R, \quad R3=\frac{R}{2}, \quad C1=C2=C, \quad C3=2C \quad \text{then } f_N = \frac{1}{2\pi RC}.$$

Gyrator

Gyrators are circuits that simulate inductors. They are typically used to simulate large inductors in audio circuits where wire wound inductors would be large, heavy, and expensive.

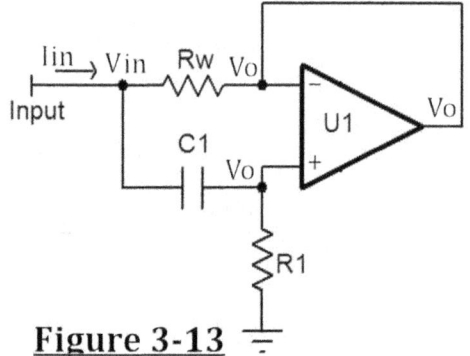

Figure 3-13

The input impedance of the gyrator circuit can be derived using a node voltage equation for the positive op-amp terminal and two approximations.

1. The voltages at the positive and negative op-amp terminals are equal.

2. The current through C1 is negligible compared to the current through R1.

$$\left(\frac{Vo-Vin}{-j/\omega C_1}\right)+\frac{Vo}{R_1}=0 \quad \text{(approximation 1)}.$$

$$Iin = \frac{Vin-Vo}{R_W} \quad \text{(approximation 2)}.$$

Solve the equations above for the ratio of the input voltage to input current to determine the gyrator's input impedance.

$$Zin=\frac{Vin}{Iin}=R_W + j\omega C_1 R_1 R_W = R_W + j\omega L_g.$$

$L_g = C_1 R_1 R_W$ where is the the effective inductance of the gyrator.

R_W is similar to wire resistance in a wire wound inductor.

Note that R_1 must be much larger than R_W for approximation 2 to be true.

The Q of the gyrator, as for a real inductor, is dependent on the frequency and on the value of Rw. The value of Rw is limited by the current capability of the op-amp. The smallest value of Rw typically used is 100Ω. The value of R1 should be greater than 10k ohms. Values above 100k ohms are typically used.

Figure 3-14 below shows an *LTspice* gyrator circuit that simulates a 22H inductor and its equivalent circuit using a real inductor with a 100 ohm winding resistance. Parallel resonance is demonstrated with a 100nF capacitor.

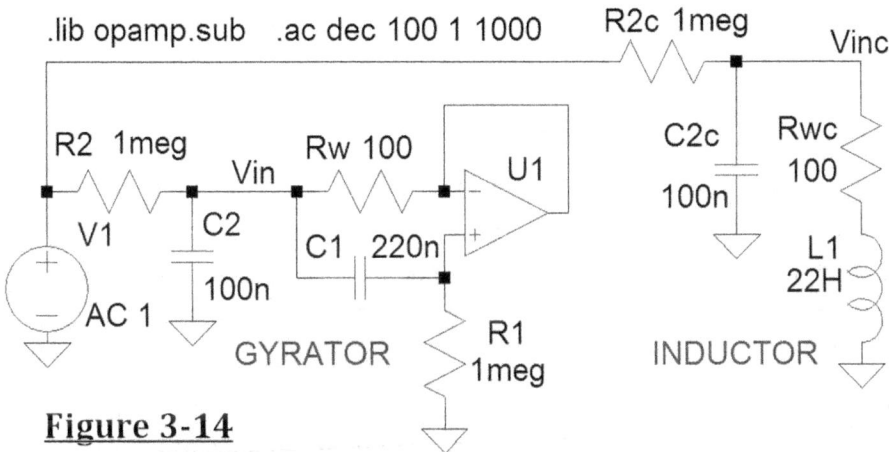

Figure 3-14

The simulation results show that the response of the gyrator circuit is almost identical to the real inductor.

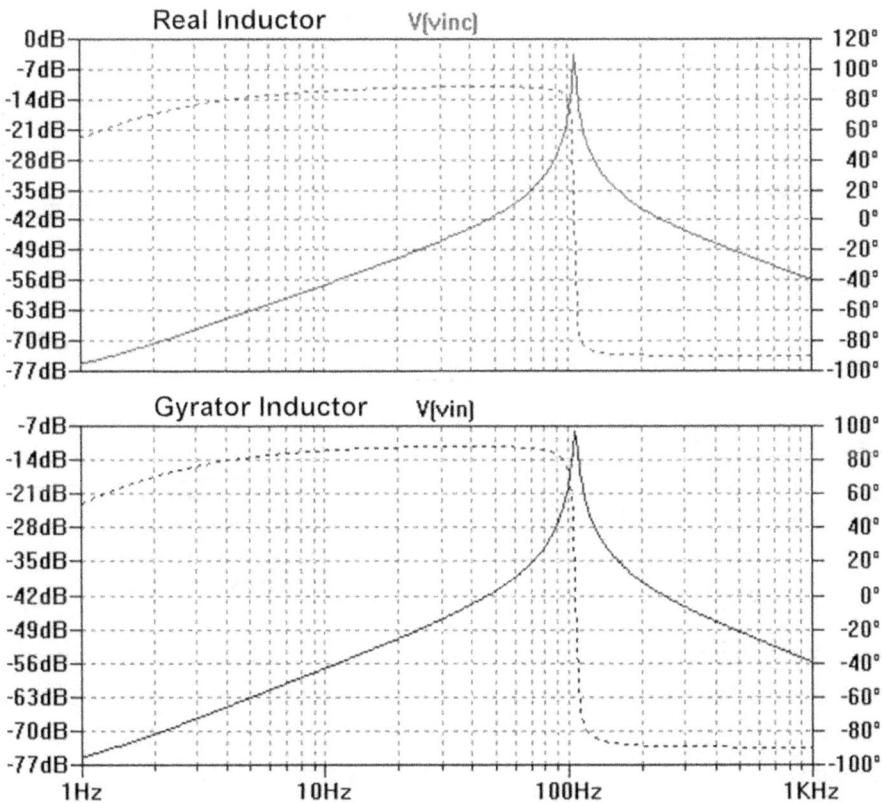

The simulation results also show that the resonant frequency of the gyrator circuit agrees with the gyrator's calculated inductance value, L_g.

$$L_g = C_1 R_1 R_W = (220 \times 10^{-9})(1 \times 10^6)100 = 22H.$$

$$\omega_n = \frac{1}{\sqrt{22(100 \times 10^{-9})}} = 674.2. \qquad f_n = \frac{\omega_n}{2\pi} = 107.3 Hz.$$

Negative Impedance Converter

A Negative impedance input means that a rise in the input voltage causes current to flow out of the input rather than into it. There are two common op-amp implementations of the negative impedance converter "NIC": current negative impedance converter "INIC" and voltage negative impedance converter "VNIC".

NIC applications include impedance matching, impedance scaling, and impedance conversion. They may a part of another circuit such as an oscillator, filter, or amplifier.

Negative resistance means that the slope of the voltage over current curve is negative instead of positive. The resistance itself is still positive. Some people consider NICs to be mysterious circuits. Negative resistance need not be so mysterious unless one thinks that it is a resistance less than 0 ohms (that would be mysterious).

INIC – Negative Impedance Converter

Figure 3-15 shows the circuit of an INIC. Its input impedance is calculated below.

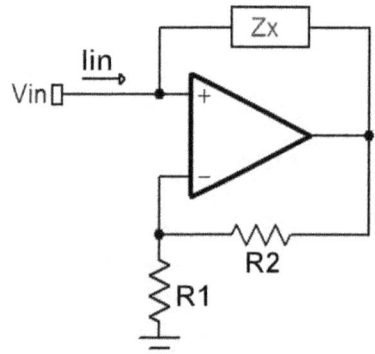

$$Vin = \frac{Vin - Vo}{Zx} \quad \Rightarrow \quad Vo = Vin - Iin Zx.$$

$$Vin = \left(\frac{R1}{R1+R2}\right) Vo = \frac{R1}{R1+R2}(Vin - Iin Zx).$$

$$Vin R1 + Vin R2 = Vin R1 - Iin Zx R1.$$

$$Zin = \frac{Vin}{Vin} = -\frac{R1}{R2} Zx.$$

Figure 3-15

Zx may be a resistance, inductive or capacitive reactance, or an impedance network. Note that the value of Zx is scaled by the ratio of R1 to R2 and its sign is reversed. R1 and R2 may also be impedances or impedance networks. Some combinations of impedances may result in an unstable circuit.

38

Figure 3-16 shows an LTspice INIC circuit using an ideal op-amp. The input voltage V1 is swept from 0 to volts 5 volts in 0.1 volts steps.

A one ohm resistor, Rc, is placed in series with V1. When plotting the current in LTspice one needs to be sure that the current arrow for Rc points to the circuit's input. If not, reverse the direction of Rc. This way the plot will show the current as positive if it is flowing in and negative if it is flowing out.

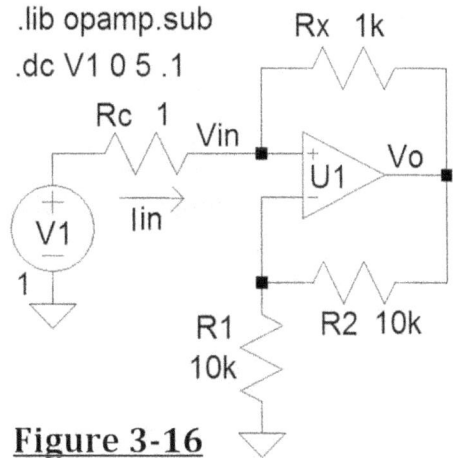

.lib opamp.sub
.dc V1 0 5 .1

Figure 3-16

The simulation result below shows the input current has a negative slope, I(Rc), decreasing from 0mA to -5mA as the input voltage increases from 0 to 5 volts. The negative slope of the current plot shows the circuit's input acts like a negative resistance. The values of the current indicate that the magnitude of the input impedance is -1000 ohms. The input impedance may be scaled up or down by changing the ratio of R1 to R2.

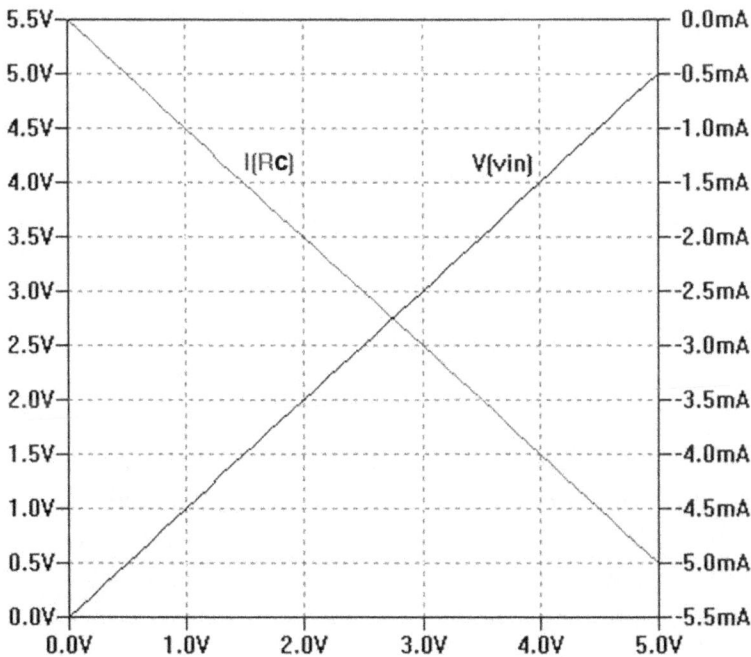

VNIC – Negative Impedance Converter

Figure 3-17 shows an *LTspice* circuit of a VNIC. A portion of the op-amp's output voltage, Vo, is fed back into the op-amp's positive input. This causes a similar change in voltage on the op-amp's negative input.

This circuit's input impedance calculation is similar to the INIC's and the result is the same.

.lib opamp.sub
.dc V1 -.05 .05 1m

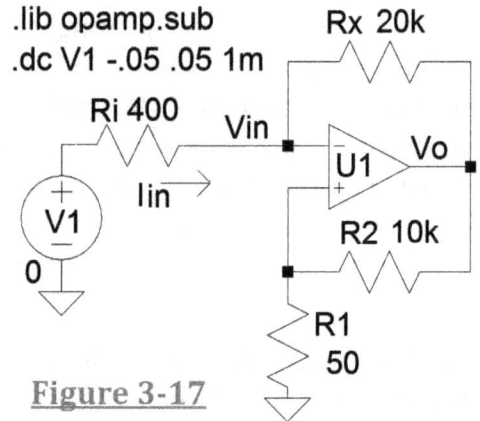

$$Iin = \frac{Vin - Vo}{Rx} \Rightarrow Vo = Vin - IinRx.$$

$$Vin = \frac{R_1}{R_1 + R_2} Vo = \frac{R_1(Vin - IinRx)}{R_1 + R_2}.$$

$$Zin = \frac{Vin}{Iin} = -\frac{R_1}{R_2} Rx.$$

Figure 3-17

Ri is the source impedance. Rx, R1, and R2 can be replaced with an impedance or impedance network. A DC sweep analysis was performed with V1 swept from -50mV to +50mV in 1mV steps. The results are displayed below.

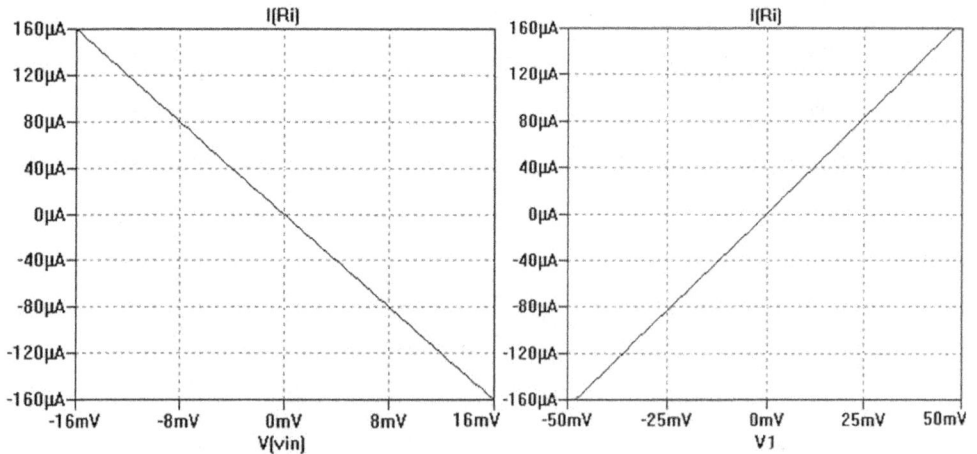

The circuit's input exhibits negative resistance since the input current is decreasing as the input voltage increases. The graph on the right shows that the input current is increasing as the input voltage increases. This is because of the 400 ohm internal impedance of the source. The source impedance is greater than the circuit input impedance, which is calculated below.

$$Zin = -\frac{R_1}{R_2} Rx = -\frac{50}{10k} 20k = -100\Omega.$$

Chapter 4: Oscillators

Two types of oscillator circuits are presented in this chapter: feedback oscillators and relaxation oscillators. Feedback oscillators can produce sinusoidal waveforms while relaxation oscillators produce square and exponential wave forms.

The operating principle of a feedback oscillator is illustrated by the block diagram on the right. The op-amp provides gain and a 180 degree phase shift. The feedback circuit provides attenuation and a 180 degree phase shift at the selected frequency. The total phase shift is 360 degrees so that the signal arrives at Vi in phase. A_V is set so that the total gain, βA_V, is one and the circuit oscillates.

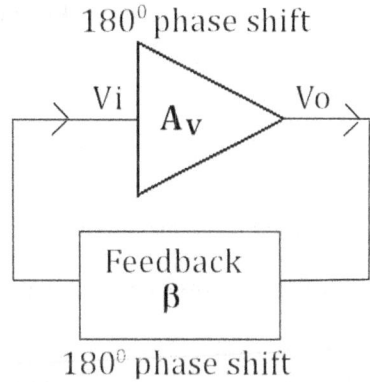

180^0 phase shift

180^0 phase shift

The operating principle of a relaxation oscillator is illustrated by the block diagram on the right. The oscillating frequency is determined by the RC constant and the voltage divider, β. The Schmitt trigger is triggered when V_{trig} crosses the value of V_{ref}. This causes Vo to change from one rail to the other so the capacitor starts charging in the opposite direction. When V_{trig} crosses the value of V_{ref} again the cycle repeats.

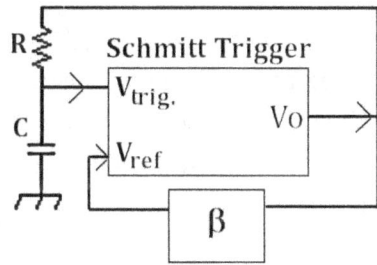

Relaxation Oscillator

Figure 4-1 is the LTspice schematic diagram of a relaxation oscillator. The circuit's time constant is R1C1. The divider R2 and R3 supply the reference voltage, V_{ref}.

The oscillator's period and frequency are given by the equations below.

$$T=2C1R1\ln\left(1+\frac{2R3}{R2}\right)=220\mu S.$$

$$f_0=\frac{1}{T}=\frac{1}{220\mu S}=4545Hz.$$

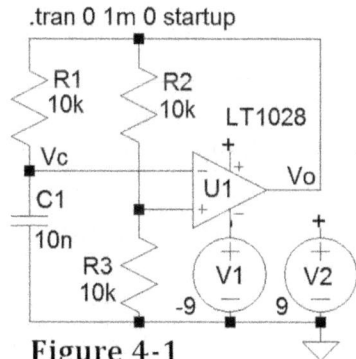

.tran 0 1m 0 startup

Figure 4-1

41

Use the real op-amp model to simulate this circuit or other oscillator circuits. The graph on the right shows the square wave output waveform Vo. The saturation voltages are determined by the op-amp model used, about ±7 volts here. The capacitor waveform, V_C, is exponential.

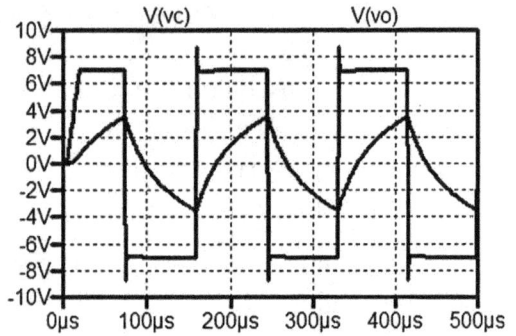

Transient analysis is used to simulate this circuit. Stop time was set to 500mS. The box "Start external DC supply voltages at 0V" must be checked to give oscillator circuits a kick start.

Triangle wave oscillator

Figure 4-2 is the LTspice schematic diagram of a triangle wave oscillator. It is still a relaxation oscillator except that the capacitor is charged by an integrator, U3. This causes the capacitor to charge linearly, thus producing a triangle wave form.

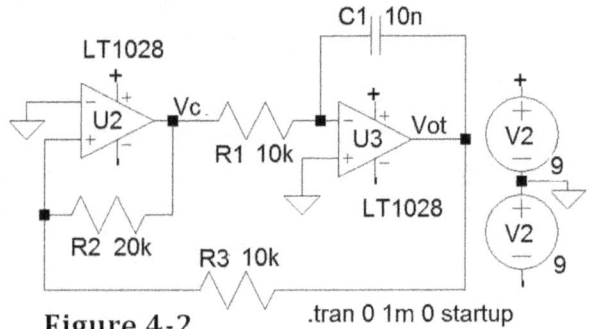

Figure 4-2

$$V_{UT} = \frac{R3}{R2}(V_{SAT+}) = \frac{10k}{20k}(7V) = 3.5V. \qquad V_{UT} = \frac{R3}{R2}(V_{SAT-}) = \frac{10k}{20k}(-7V) = -3.5V.$$

$$Vo_{PP} = V_{UT} - V_{LT} = 3.5V - (-3.5V) = 7V_{PP}.$$

$$f_0 = \frac{1}{2R1C1}\frac{R2}{R3} = \frac{1}{2(10k)(10n)}\frac{20k}{10k} = 5000Hz \quad \Rightarrow \quad T = 200\mu s.$$

The simulation result on the right shows that the period is about 200μS, and the peak-to-peak amplitude is 7V, as expected.

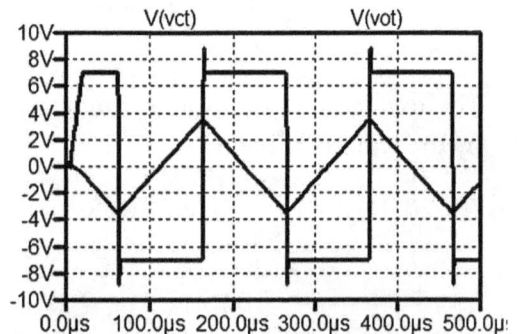

42

Phase Shift Oscillator

Figure 4-4 shows an LTspice schematic diagram of an one op-amp phase shift oscillator. The RC network provides a 180 degree phase shift at the oscillating frequency. It also introduces an attenuation of 29. The gain of the op-amp must be at least 29 to compensate for the attenuation of the filter.

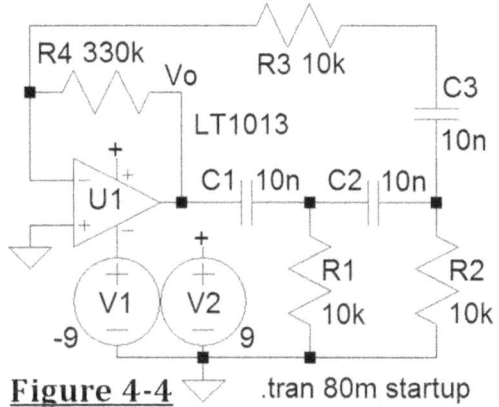

Figure 4-4

$$R=R1=R2=R3=10k, \quad C=C1=C2=C3=10nF.$$

$$f_0 = \frac{1}{2\pi RC\sqrt{6}} = \frac{1}{2\pi 10k\,10n\sqrt{6}} = 650Hz. \quad R4>29R.$$

Sinusoidal oscillator circuits require a start-up time to reach steady-state. Also the "Start external DC supply voltages at 0V" in the transient analysis window must be checked. In this case a transient simulation was performed for 80mS. The simulation result below shows that it took 70mS for this circuit to reach steady-state.

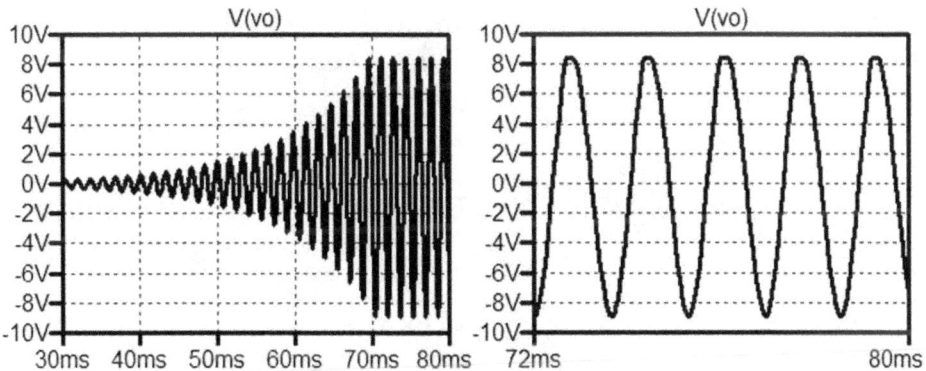

The expanded graph above on the right shows a steady-state sinusoidal wave form starting at 72mS. This circuit was simulated with R4 set to 330k instead of the theoretical value of 290k. The extra gain improves reliability, but too much gain would cause the peaks of the sine wave to flatten.

The period of the simulation is about 1.6mS: $T=1.6mS$, $f_0 = \frac{1}{T} = 625Hz$.

43

3-Phase Oscillator

Three op-amps in a phase shift oscillator circuit can be used to generate three phase voltages. Each op-amp filter is designed to produce a 120 degree phase shift at the oscillating frequency. Refer to the transfer function of an op-amp single-pole low-pass filter.

$$\left|\frac{V_o}{V_s}\right| = \frac{-A_{INV}}{\sqrt{1+\left(\omega/\omega_c\right)^2}}, \qquad \omega_c = \frac{1}{RC}, \qquad \theta = 180 - \arctan\frac{\omega}{\omega_c}.$$

Refer to the *LTspice* circuit in figure 4-5 below. Since each op-amp acts as a buffer, the transfer function for the entire circuit is the product of the transfer functions of each stage.

$$\theta(\text{each stage}) = 120^0 = 180^0 - \arctan\frac{\omega}{\omega_c} \;\Rightarrow\; \arctan\frac{\omega}{\omega_c} = 180^0 - 120^0 = 60^0.$$

$$\frac{\omega}{\omega_c} = \tan 60^0 = 1.732. \quad \text{If } \omega = 2\pi 400 = 2513 r/s, \quad \omega_c = \frac{2513 r/s}{1.732} = 1451 r/s.$$

$$\left|\frac{V_o}{V_s}\right| = \frac{-A_{INV}}{\sqrt{1+(1.732)^2}} = -1 \;\Rightarrow\; A_{INV} = \sqrt{1+(1.732)^2} = 2.$$

For the circuit to oscillate at $2513 r/s$, the cutoff frequency of each stage must be $1451 r/s$. If we choose C = 39nF and stage gain =2.37.

$$\omega_c = \frac{1}{RC} \;\Rightarrow\; 1451 r/s = \frac{1}{R(39nF)} \;\Rightarrow\; R = \frac{1}{(1451 r/s)(39nF)} = 17.7k.$$

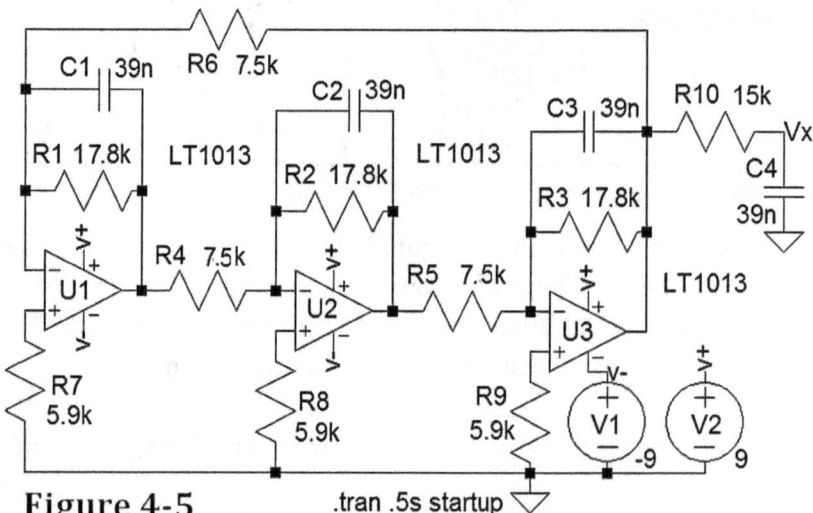

Figure 4-5   .tran .5s startup

44

The transient analysis stop time was set to 500mS. Check the "Start external DC supply voltages at 0V" box. It took 400mS for the circuit to reach steady-state.

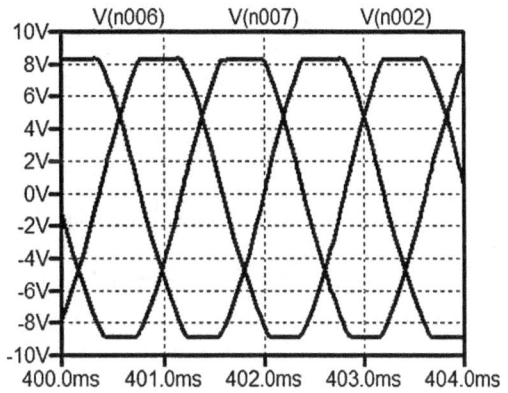

The graph on the right shows the three-phase waveforms with flattened peaks. Reducing the gain of each stage closer to 2 would improve the sine waves, but the startup time would be longer and the circuit may be less stable.

A low-pass filter was added to the third stage to filter the flat-peaked sine wave. The result is shown on the right. This filter does add a phase shift, but if the same filter is used for each stage, the output phases would still be 120⁰ apart.

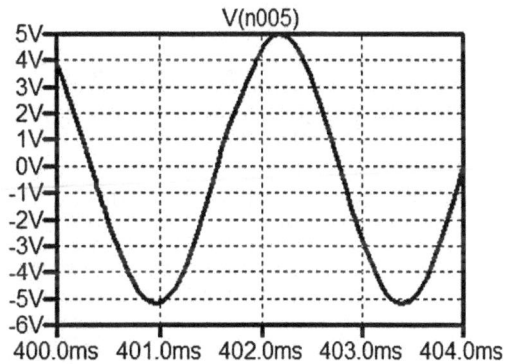

Wien Bridge Oscillator

Figure 4-6 shows an LTspice schematic diagram of a one op-amp Wien bridge oscillator. This circuit oscillates at the frequency where Vi is in phase with Vo and the loop gain equals 1.

This is a very popular analog sine wave oscillator which was first implemented with vacuum tubes and transistors.

Typically the analysis is simplified by setting R4 = R3 = R and C1 = C2 = C. The node voltage method is used below.

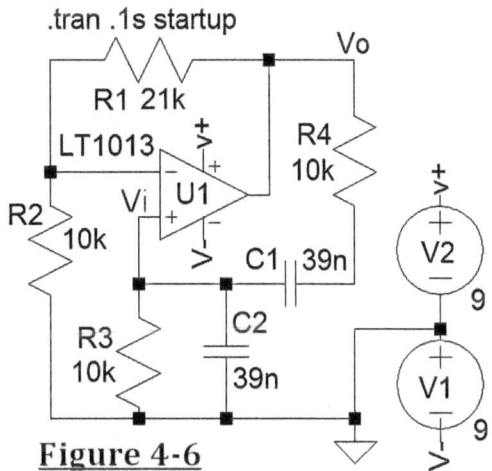

Figure 4-6

$$\frac{V - V_o}{R - j\frac{1}{\omega C}} + \frac{V}{R} + \frac{V}{-j\frac{1}{\omega C}} = 0 \quad \Rightarrow \quad \frac{j\omega C(V - V_o)}{j\omega CR + 1} + \frac{V}{R} + j\omega CV = 0.$$

45

Solving for $\dfrac{\text{Vo}}{\text{V}}$: $\dfrac{\text{Vo}}{\text{V}} = \dfrac{j\omega CR + (1+j\omega CR)^2}{j\omega CR} = 3 + j\left(\omega CR - \dfrac{1}{\omega CR}\right).$

Phase angle is 0 when $\omega CR = \dfrac{1}{\omega CR} \Rightarrow \omega = \dfrac{1}{RC}.$

When phase angle is 0, Voltage gain=3.

Voltage gain for this circuit is: $A_V = 1 + \dfrac{R1}{R2} = 3.$

R1 must be equal to 2R2 theoretically, but the gain is usually set a little higher. There are a variety of practical designs that provide higher gain at start up, and a gain closer to 3 when running.

The simulation result is shown below. Transient analysis was run for 100ms. It took nearly 60mS to reach steady-state. The period displayed is 2.5ms which corresponds to a frequency of 400Hz. The voltage gain is 3.1.

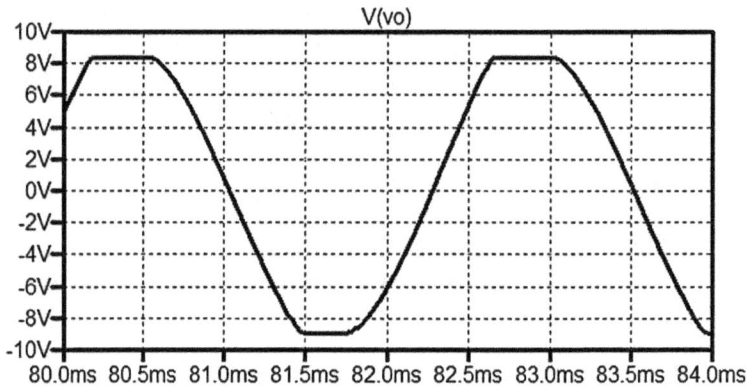
V(vo)

Theoretical frequency: $f_o = \dfrac{1}{2\pi RC} = \dfrac{1}{2\pi 10k(39n)} = 408Hz.$

46

Chapter 5: Projects

The projects in this chapter and chapters 6, 7, and 8 are intended to present basic design concepts and example applications of op-amps. Each project includes design principles, simulation information, and suggested experiments. These projects may provide ideas for other projects. Experimentation is encouraged.

Most experiments may be performed on a solder-less breadboard. Parts for the experiments are available at *Mouser Electronics* and/or *Digi-key*. A list of part sources and selected parts data is provided in the appendix. A bi-polar DC power supply and a multi-meter are sufficient for some experiments while others will also require an oscilloscope and function generator.

Simulation Software

Simulation is an important part of the design and testing process. A variety of SPICE based simulation software is available from free to very expensive.

Most simulation examples use *LTspice* which is available as a free download from *Linear Technology*. There is no limit on the number of nodes or on library parts. *www.linear.com*.

Another free simulator, *TINA-TI*, is available from *Texas Instruments*. This is a version of *TINA* (Toolkit for Interactive Network Analysis). *TI* quote: "TINA-TI is ideal for designing, testing, and troubleshooting a broad variety of basic and advanced circuits, including complex architectures, without any node or number of device limitations". "This complimentary version, TINA-TI, is fully functional but does not support some other features available with the full version of TINA". Texas Instruments: *www.ti.com*.

TINA is available from *Design Soft*: *www.tina.com*.

Cadence Design Systems offers a free demo version of their software, *PSpice*, with a limited number of nodes a limited parts library. *www.cadence.com*

One consideration in choosing simulation software is the software's parts library. For example, *LTspice* has a complete library of *Linear Technology's* op-amps while *TINA-TI* has a complete library of *TI* op-amps. However, both programs allow the importation of third party models.

Design Soft's TINA has a large selection of parts from a variety of manufacturers. There are several purchase options. The basic program sells for $129 (2014 price) and is limited to 200 nodes. However there are no other limits on parts or functionality. This program does have some novel features.

Project 1: Variable Voltage Power supply

This type of voltage regulator may be used to derive a lower voltage from a circuit's main power supply. Its advantages include simplicity and lack of switching noise. However, because of its low efficiency this design is mainly used for low currents. This project is about the design, simulation, building and testing a linear variable voltage power supply with the following specifications.

1. Output voltage is variable from 2V to 9V with good load regulation.
2. Input 12VDC regulated. Maximum current: 100mA.

Design Principles

The circuit in figure 5-1 is an application of a voltage controlled voltage source. The control voltage is varied by the potentiometer and applied to the op-amp's non-inverting terminal. The op-amp's output current is amplified by the transistor, Q1. Both U1 and Q1 have unity voltage gains, and large current gains.

U1 uses a single 12V power supply. Its positive rail is connected to +12V and its negative rail is connected to ground.

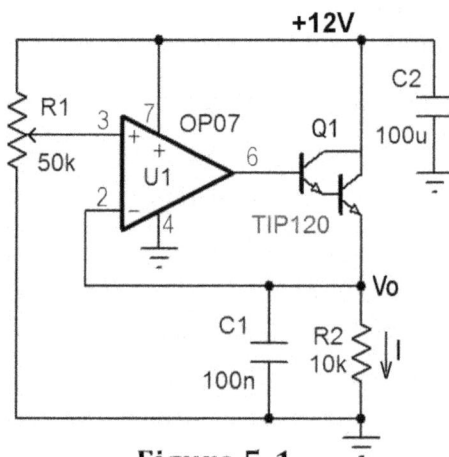

Figure 5-1

The op-amp's output must be able to supply Q1's base current. The amount of current required depends on the transistor's current amplification factor, β, and its emitter current. In this case, the maximum emitter current is 100mA. A transistor with a current gain of 50 would require a maximum base current of 2mA.

If the power supply were designed to supply 1000mA, the transistor's base current would be 20mA, too high for the op-amp. However, a high current gain darlington transistor could be used. A transistor with a β of 1000 would require base current of only 1mA to supply an emitter current of 1000mA.

The power dissipation of the transistor must also be considered. For example, when the power supply output voltage is 4 volts and its output current is 100mA, the transistor will disspate 800mW.

$$P_Q = (V_C - V_O)I_C = (12V - 4V)100mA = 800mW.$$

48

Simulation

The *LTspice* circuit diagram for this project is presented in figure 5-2.

Output voltage, Vo, is varied by the voltage divider, R1 and R2. R1 and R2 simulate the potentiometer. A parametric sweep is used to vary the values of R1 and R2.

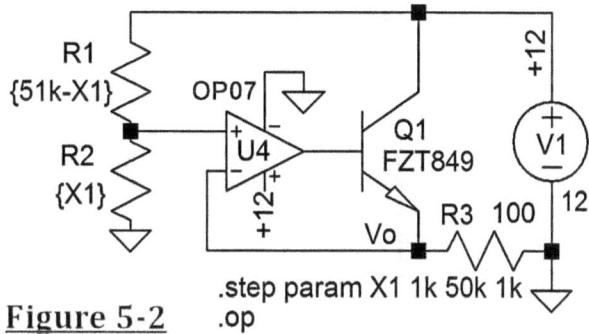

Figure 5-2

.step param X1 1k 50k 1k
.op

The value of R2 is given by the parameter {X1}. X1 must be enclosed in curly brackets as shown. The value of R1 is given by the expression {51k-X1}, and it must also be enclosed in curly brackets as shown.

This simulation introduces parametric sweep analysis, which may be used to vary the value of any part. Alternatively, R1 and R2 could be replaced by a DC power supply and DC sweep analysis could be used to vary its voltage from 0V to 12V.

The value of X1 is varied from 1k to 50k in 1k steps using the ".*step*" command. Enter the command by selecting "*SPICE Directive*" from the "*Edit*" menu. Type the command into the dialog box: .*step param X1 1k 50k 1k*. For more information consult *LTspice Help* on the ".*step*" command.

Set the analysis type to "*DC op pt*". Run the simulation. Use *LTspice's* voltage probe to graph the output voltage, Vo, at the transistor's emitter as function of the value of the parameter X1. Use *LTspice's* current probe to graph the current flowing into the transistor's base as function of the value of the parameter X1.

In addition to the saturation voltages of the op-amp, the output voltage range is limited by the transistor's saturation voltage and base to emitter voltage drop. A power transistor will typically have a base to emitter voltage drop of 0.7V to 1V. A Darlington transistor such as a TIP120 may have a base to emitter voltage drop of about 1.4V to 2V. Refer to the appendix for TIP120 data.

Experiment with other op-amps, such as an LT1013 or an OP07. See how the op-amp's rails effect the range of Vo. Refer to the appendix for a data table of a selection of op-amps.

49

The results of a DC operating point analysis "*.op*" are presented below.

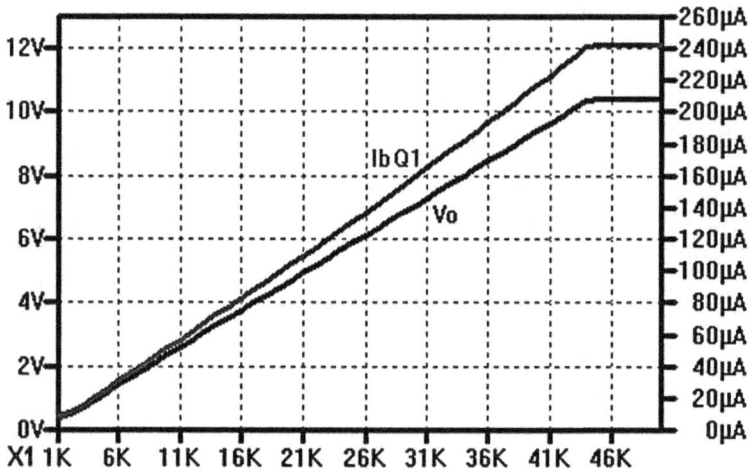

The graphs show the output voltage Vo at the transistor's emitter and the current into the transistor's base as functions of the value of the parameter X1. The maximum output voltage is about 10.4V. The transistor's base current, Ib, is also plotted. The potentiometer voltage (voltage across R2) could also be plotted as a function of the parameter X1.

Change transistors by right clicking on the transistor. A dialog box will present an assortment of transistors with SPICE parameters included. The FZT849 was used for this simulation.

Parts Selection

The value of the potentiometer is not critical. However low values will draw excessive currents and high values may not supply a stable voltage. A 50k pot is chosen here. Its current is 0.24mA and its power dissipation is 2.9mW.

Any general purpose op-amp may be used, such as the OP07. A TIP120 power transistor (Darlington) is chosen for Q1. Any similar transistor may be used. Q1 supplies maximum power to the load when the load is 90Ω and the output voltage is 9 volts. P_{load} = 9V(100mA) = 900mW. P_{Q1} = 3V(100mA) = 300mW.

The maximum power dissipation for Q1 is calculated below.

$$P_{Q1} = (12 - Vo)\frac{Vo}{R_L} = \frac{12Vo}{R_L} - \frac{Vo^2}{R_L}, \quad \text{Max power occurs when} \frac{dP_{Q1}}{dVo} = 0.$$

$$\frac{dP_{Q1}}{dVo} = \frac{12}{R_L} - \frac{2Vo}{R_L} = 0 \implies Vo = 6V, \quad P_{Q1(max)} = (12V - 6V)100mA = 600mW.$$

Experiment

Parts

Resistors: 50k pot.   100Ω, 2W, 5%.   10k, ¼W, 5%.
Capacitors: 2-100nF.   Q1: TIP120 or equivalent. U1: OP07CP or equivalent.

Procedure

1. Build the circuit in figure 5-1 on a breadboard. Use a trim-pot for the 50k potentiometer. Refer to appendix 2 for more information on the TIP120.

2. Apply exactly 12.0VDC to the circuit and verify that the pot will vary the output voltage from 0V to at least 9V.

3. Set the output voltage to exactly 6.0V. Verify that the input voltage is still exactly 12.0V. Connect a 100Ω, 2W resistor to the output. Verify that the input voltage is still exactly 12.0V. Measure and record the resulting output voltage, Vo. Disconnect the resistor after making the measurement (it will get warm).

4. Calculate the circuit's load regulation. %Reg = 100 (6 –Vo)/Vo.

Analysis

1. The voltage on the non-inverting input of the op-amp will be very close to the output voltage Vo for this circuit. This circuit's load regulation is dependent on the stabilty of the voltage on pin 3 of the op-amp. This voltage will change if a change in load causes the 12 volt supply voltage, V1, to change.

   This circuit has poor line regulaton because any change in the line voltage will result in a change of voltage on the op-amp's non-inverting input. This will result in a change of Vo.

   Calculate the theoretical value of this circuit's line regulation for a change in V1 from 12.0V to 14.0V. Hint: Vo is 6.0V when V1 is 12.0V. The ratio of R1 to R2 is constant. What is Vo when V1 is 14.0V?

2. Calculate the expected temperature of the transistor when it's dissipating maximum power. Refer to data in appendix 2 if you used the TIP120. The TIP120 can dissipate much more power when used with a heat sink. However, due to the poor efficiency of the linear regulator, switching regulators are often preferred for higher powers.

3. Calculate the efficiency of this circuit in supplying power to a 100 ohm load for Vo = 3V and Vo = 9V.

Project 2: Regulated Power Supply

This project is similar to project 1 with two exceptions. First, the reference voltage is provided by an accurate voltage reference. Second, the reference voltage is amplified by the voltage gain of U1. The output voltage is adjustable by varying the gain of U1. This project is about the design, simulation, building and testing a linear variable voltage power supply with the following specifications.

1. Output voltage is continuously variable from 1.5V to 9V.
2. Input 12VDC. Maximum current: 100mA.
3. Good load and line voltage regulation.

Design Principles

Refer to figure 5-3 below. The voltage on the op-amp's non-inverting input is held constant by the 1.2V voltage reference IC, U2 (symbolized as a zener diode) to improve the power supplies line and load voltage regulation. Voltage references are available in a variety of voltage values and accuracies.

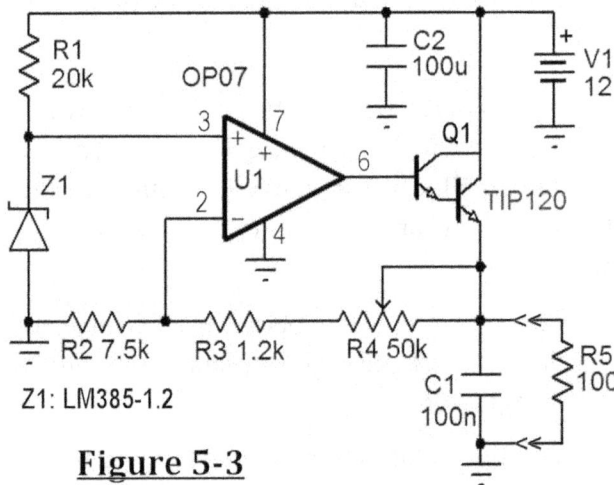

Figure 5-3

The op-amp circuit is a non-inverting amplifier whose output current is amplified by the transistor, Q1. This transistor is configured as an "emitter follower". It has unity voltage gain and a large current gain.

The regulator's output voltage, Vo, depends on the voltage gain of the op-amp.

$$Vo = \left(1 + \frac{Rf}{Ri}\right) Vref = \left(1 + \frac{Rf}{7.5k}\right) 1.25. \qquad 1.2k \le Rf \le 51.2k.$$

Vo=1.45V when Rf=1.2k.   Vo=9.8V when Rf=51.2k.

Simulation

The *LTspice* circuit diagram for this project is presented in figure 5-4 on the right. It is a voltage follower circuit whose output current is amplified by the transistor, Q1.

A *Linear Technology* LT1004-1.2 is used for the voltage reference in the simulation. This is a precision voltage reference that is similar to the LM385-1.2, but it is considerably more expensive. than the LM355

DC sweep analysis was used to simulate the circuit. The supply voltage was swept from 11V to 15V. The result is shown on the right.

The %Reg calculation below shows that the line regulation using the LT1004-1.2 is very good.

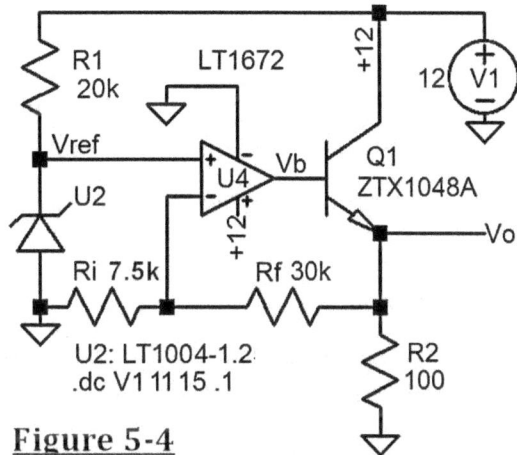

Figure 5-4

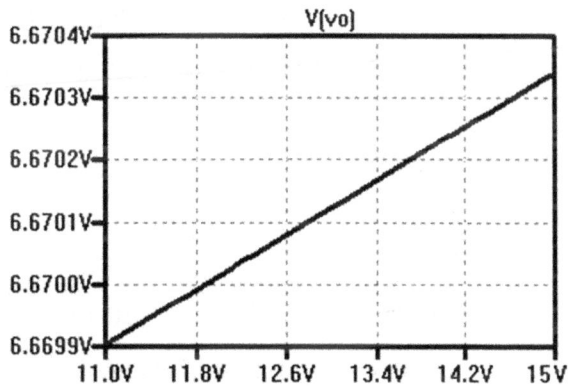

$$\% Reg = \frac{6.66990 - 6.67034}{6.6699} \cdot 100\% = .0066\%.$$

$$\Delta V1 = 4V. \qquad \% Reg / V = .00165\% / V.$$

Part Selection

Any general purpose op-amp may be used, such as the OP07. The transistor, Q1, supplies maximum power when the load is 90Ω and the output voltage is 9 volts. $P_{load} = 9V(100mA) = 900mW$. $P_{Q1} = 3V(100mA) = 300mW$.

Q1 dissipates maximum power when the output voltage is 6 volts and output current is 100mA.. $P_{load} = 6V(100mA) = 600mW$. $P_{Q1} = 6V(100mA) = 600mW$.

A TIP120 power transistor is chosen for Q1. Any similar transistor may be used. Since the op-amp's reference voltage is regulated, the circuit's supply voltage does not need be regulated. A common and inexpensive voltage reference, the LM385-1.2, is used for this project.

53

Experiment

Parts

Resistors: 50k pot. 100Ω, 2W, 5%. 2k, 7.5k, 20k, ¼W, 5%.
Capacitors: 2-100nF. Q1: TIP120 or equivalent.
U1: OP07CP or equivalent. U2: LM385-1.2

Construction and testing

1. Build the circuit in figure 5-3 on a breadboard. Use a trim-pot for the 50k potentiometer.

2. Apply 12VDC to the circuit and verify that the pot will vary the output voltage from 1.5V to at least 9V.

3. Set the output voltage to 9.0V. Connect a 100Ω, 2W, resistor to the output. Measure and record the resulting output voltage, V_{O1}. _____.

   Increase the circuit's power supply voltage from 12 volts to 15 volts. Measure and record the resulting output voltage, V_{O2}. _____.

   Turn off the power after making the measurements (*warning*: the resistor and transistor will get hot).

Analysis

1. Calculate the circuit's load regulation.

2. Calculate the "worst case" value of the load regulation from the LM385-1.2 data sheet.

$$\left(\%Reg = \frac{(9 - V_{O_1})100\%}{V_{O_1}}. \right)$$

3. The quality of line voltage regulation is often expressed in percent per volt change of input voltage (%/V).

$$\left(\%/V = \frac{100(V_{O_2} - V_{O_1})}{(15 - 12)^2}. \right)$$

4. Calculate the "worst case" value of the line regulation from the data sheet for the LM385-1.2 (refer to the appendix).

5. Compare your experimental results to the calculated "worst case" values.

Project 3: Light Meter and Switch

This project uses a photo-transistor to sense the intensity of visible light. The circuit can be used as a photometer and as a light activated relay.

Design Principles

The emitter current of Q1 in figure 5-5 below is proportional to light intensity. The resulting voltage drop across R1 is amplified by U1A. Pin 1 of U1A can be connected to a voltmeter to read the light intensity.

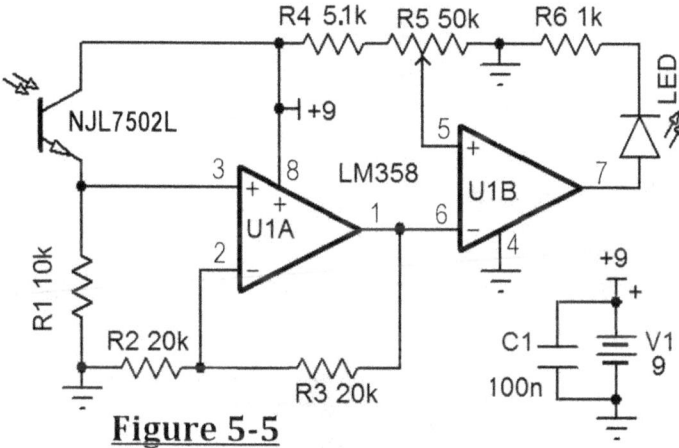

Figure 5-5

U1B is an analog comparator. When the voltage on pin 6 is less than the voltage on pin 5, the output voltage of U1B is its positive saturation voltage. When the voltage on pin 6 is greater than the voltage on pin 5, the output voltage of U1B is its negative saturation voltage. The voltage on pin 5 is adjustable with R3. D1 is an LED indicator that lights when pin 7 of U1B is at its positive saturation voltage.

The characteristics of the photo-transistor, Q1, and the value of R1 determines the voltage applied to U1A pin 3 for a particular light intensity.

The graph on the right shows the response of an NJL7502L transistor. Its spectral response is similar to that of a human eye.

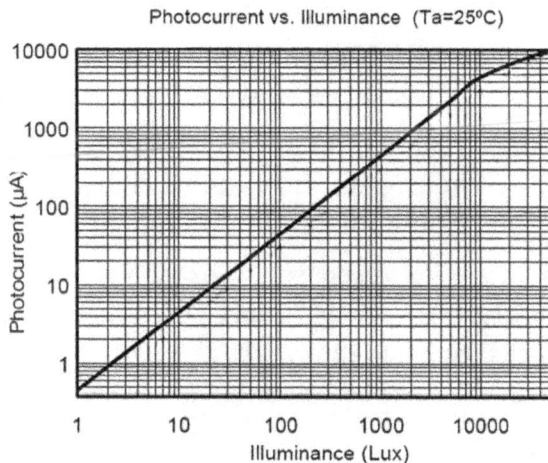

Photocurrent vs. Illuminance (Ta=25°C)

Light intensity is given in units of "Lux". Interior room lighting is typically in the range of 50 Lux to 500 Lux. Direct sunlight is in the range of 30,000 to 100,000 Lux.

Simulation

Figure 5-6 shows the *LTspice* circuit. The current source I1 simulates the photo-transistor, Q1. R1 is set to 10k and the gain of U1 is set to 2.

I1 is swept from 0µA to 500µA. The theoretical value of the output voltage of U1 is given by:

$$Vo = I1\,R1\left(1 + \frac{R3}{R2}\right) = 20,000(I1).$$

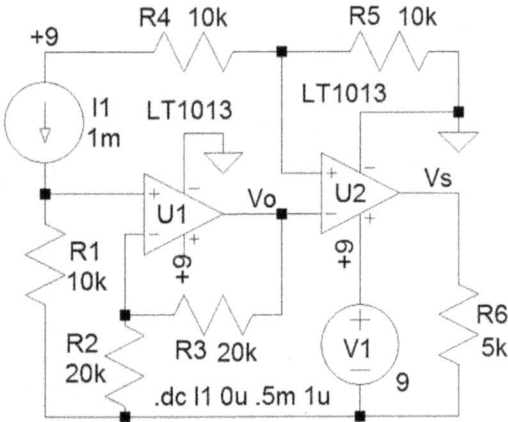

Figure 5-6

This equation predicts an output voltage of 10 volts when I1 = 500µA. The op-amp will saturate, of course, since the supply voltage is only 9V.

Simulation results for Vo and Vs are presented in the graph on the right.

The saturation voltage of U1 is shown to be about 8.3 volts. U2 is set to trigger at 4.5 volts.

Vs = 8.2V when I1<225µA.
Vs = 0V when I1>225µA.

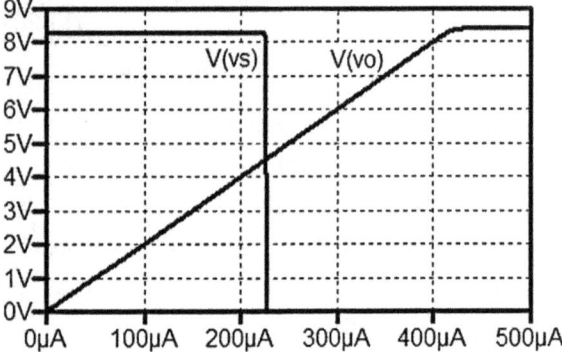

For the NJL7502L, 225µA corresponds to a light intensity of about 450 Lux. 400µA corresponds to a light intensity of about 1000 Lux.

This circuit can be set to trigger at light levels between 0 Lux and 400 Lux. The trigger range can be changed by changing the value of R1 or by changing the gain of the op-amp.

The spectral response of this circuit can be changed by changing the photo-detector. It could be changed to infrared and used as an intrusion detector or a digital communicator. The response time of this circuit is limited by the rise and fall times of the detector and the op-amp's bandwidth and slew rate.

Experiment

Parts (figure 5-5)

Q1: NJL7502L.   D1: LED.   U1: LM358 or equivalent.
R1: 10k, ¼ watt, 5%.   Ri, Rf: 22k, ¼ watt, 5%.   R2, R4: 1k, ¼ watt, 5%.
R3: 10k trim pot.   C1: 100μF, 16V.

Construction and Testing

1. Build the circuit in figure 5-5 on a breadboard using the part values given. above.

2. Verify that the voltage on pin 5 of U1B can be varied from 0V to about 8V with potentiometer R3. Set the voltage on U1B pin 5 to 4V.

3. Connect a voltmeter to pin 1 of U1A. Vary the light intensity on Q1 while observing the voltmeter.   Verify that the LED turns on and off when the voltage on pin 1 of U1A crosses 4.0V.

4. Experiment with the value of R1 (4.7k, 20k, for example. Compare the circuit's response for different values of R1.

LED's have relatively short rise and fall times, typically less than 5ns. They can be used to test the rise and fall times of the photo-detector and the time response of the circuit. A function generator with a 50 ohm output impedance and rise and fall times shorter than 100ns is required.

1. Connect the LED in series with a 220 ohm resistor to the function generator. Connect channel 1 of an oscilloscope to the output of the function generator. Set the generator to produce a 10kHz, 5V p-p, square wave with a 2.5V offset (so it goes between 0V and 5V).

2. Couple the LED to the photo-detector with a small cardboard tube or equivalent.   The LED should be about 0.5 inches from the detector. Minimize any stray light on the photodiode.

3. Connect channel 2 of the oscilloscope to R1. Set the oscilloscope to 10μs per division initially to observe the rise and fall times.   Adjust the oscilloscope amplitude and time per division for the most accurate measurement.

4. Repeat step 3 with channel 2 of the oscilloscope connected to Vo.

5. Repeat step 3 with channel 2 of the oscilloscope connected to Vs.

6. Compare the overall response times of the measurements in steps 3 to 5 above.

Project 4: Phase Tripler

It is possible to do a variety of 3-phase experiments using a low voltage three-phase source. Experiments can be done safely without the need of a three-phase power outlet. This project is about the design, simulation, building, and testing a single phase to 3 phase converter with the following specifications.

1. Input: 0 to 18V p-p amplitude single phase 60Hz sine wave.
2. Output: 0 to 18V p-p amplitude 3-phase 60Hz sine waves.
3. Supply up to 125mW per phase.

Design Principles

This is a simplified version of the ZAP Studio "Phase Tripler". It converts a 60Hz sine wave to 3-phase 60Hz sine waves. The input voltage can range from 0 to 18V p-p. The output voltage of each phase is equal to the single phase input voltage. Figure 5-7 below shows a block diagram of the phase tripler.

The input is buffered by op-amp, U1A, and output at P0. U1B shifts the input phase by 180⁰.

The phase shifted voltage is applied to a 60⁰ lag network to obtain a net phase shift of 120⁰, and to a 60⁰ lead network to obtain a net phase shift of 240⁰.

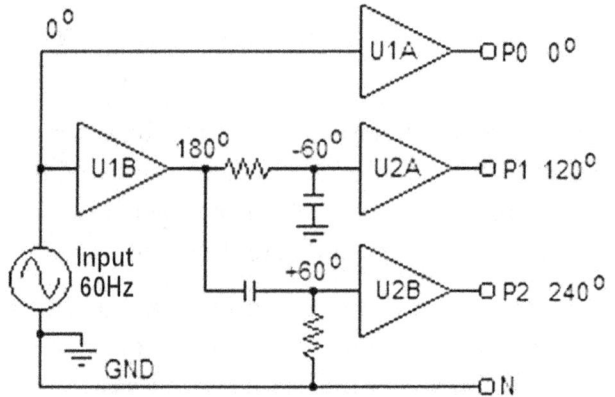

Figure 5-7

It can be shown that the output amplitude of the lag and lead networks is exactly one half of the input amplitude. To compensate, U2A and U2B have a voltage gain of two. The lag network (low pass filter) calculations are given below.

$$\frac{V_O}{V_{IN}} = \frac{-jX_C}{R - jX_C} = \frac{1}{1 + \dfrac{R}{-jX_C}} = \frac{1}{1 + j\omega RC}. \qquad \theta = -arctan(\omega RC) \Rightarrow \tan\theta = \omega RC.$$

For 60^0, 60Hz, C=56nF : $\tan 60^0 = 1.732 = 2\pi 60R\left(56\times10^{-9}\right) \Rightarrow R = 82k$.

$$\left|\frac{V_O}{V_{IN}}\right| = \frac{1}{\sqrt{\left(1^2 + (\omega RC)^2\right)}} = \frac{1}{\sqrt{\left(1^2 + 1.732^2\right)}} = 0.5.$$

The lead network (high pass filter) calculations are given below.

$$\frac{V_O}{V_{IN}}=\frac{R}{R-jX_C}=\frac{1}{1-j\frac{X_C}{R}}=\frac{1}{1-j\frac{1}{\omega RC}}. \qquad \theta=arctan\left(\frac{1}{\omega RC}\right)\Rightarrow \tan\theta=\frac{1}{\omega RC}.$$

For 60^0, 60Hz, $C=56nF$: $\tan 60^0 = 1.732 = \dfrac{1}{2\pi 60 R\left(56\times 10^{-9}\right)} \Rightarrow R=27.35k.$

$$\left|\frac{V_O}{V_{IN}}\right|=\frac{1}{\sqrt{\left(1^2+\left(\frac{1}{\omega RC}\right)^2\right)}}=\frac{1}{\sqrt{\left(1^2+1.732^2\right)}}=0.5.$$

Simulation

Figure 5-8 below shows the *LTspice* circuit diagram of the phase tripler. R4 and C1 comprise the lag network. R7 and C2 comprise the lead network. The circuit in figure 5-8 uses the part "opamp" from *LTspice's* opamp library. This part does not require power supply connections on the schematic. The directive: ".*lib opamp.sub*" must be included.

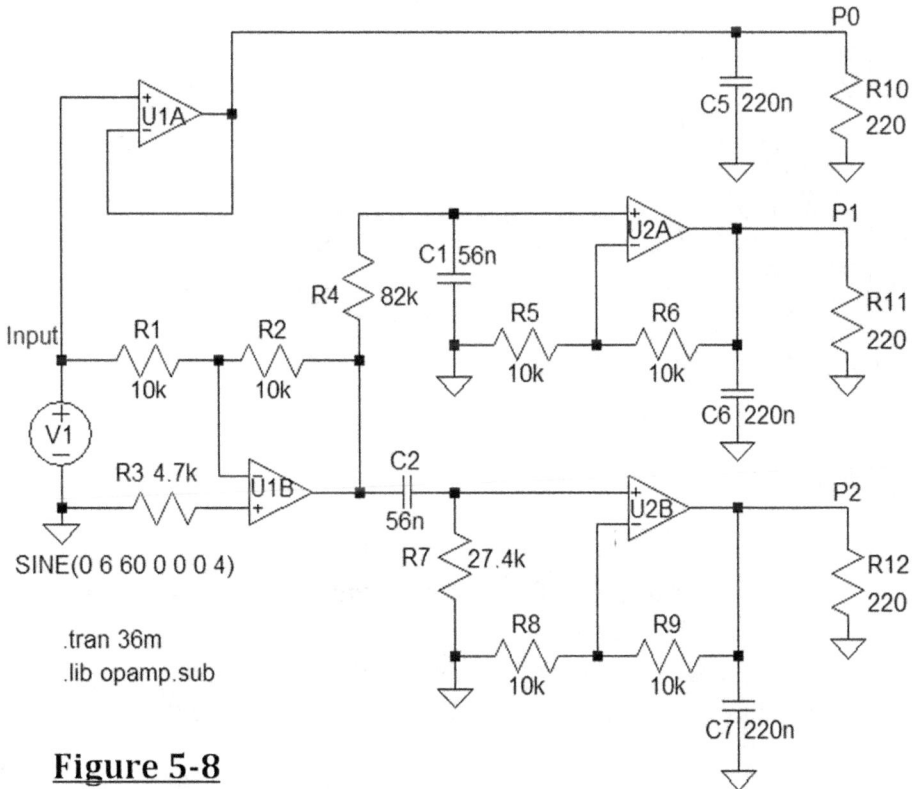

Figure 5-8

The op-amp labels correspond to the block diagram in figure 5-7.

To add a directive, click on ".op" (on the far right in the main menu bar). See below:

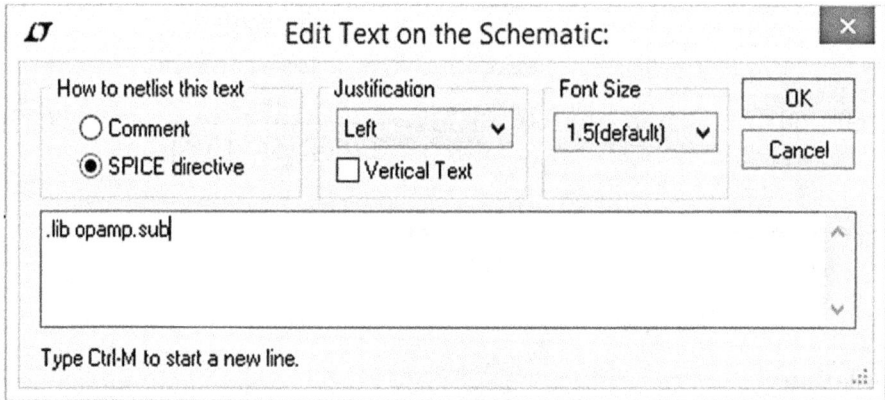

V1 is set to produce 4 cycles of a 12V p-p, 60Hz, sine wave. The simulation is set to "*Transient*" with a stop time of 36mS. The simulation result is shown below.

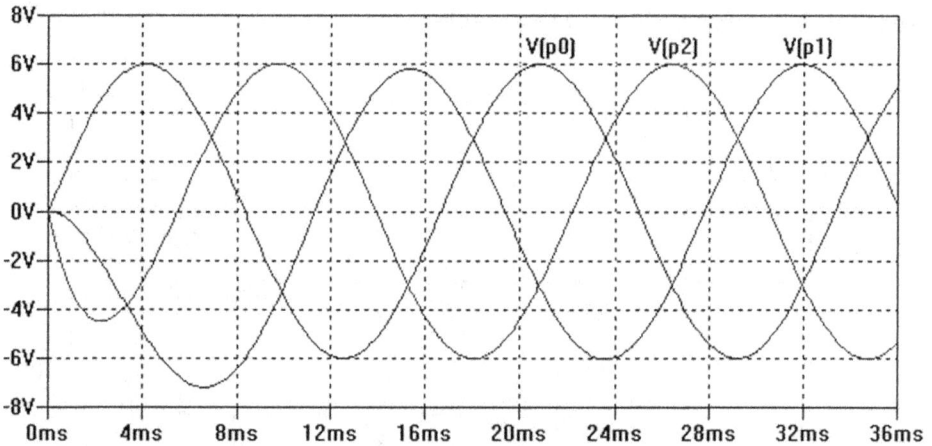

Experiment

Parts

U1, U2: L272M Power Op-amp. Observe pin numbers. Do not substitute.
R1, R2, R5, R6, R8, R9: 10k, ¼ watt, 1%.
R3: 4.7k, ¼ watt, 5%.   R4: 82.0k, ¼ watt, 1%.   R7: 27.4k, ¼ watt, 1%.
C1, C2: 56nF, 2%.   C3, C4: 100µF, 25V.   C5, C6, C7: 220nF, 10%.

Construction and testing

1. Build the circuit in figure 5-9 on a breadboard using the part values given above. V1 is a function generator.

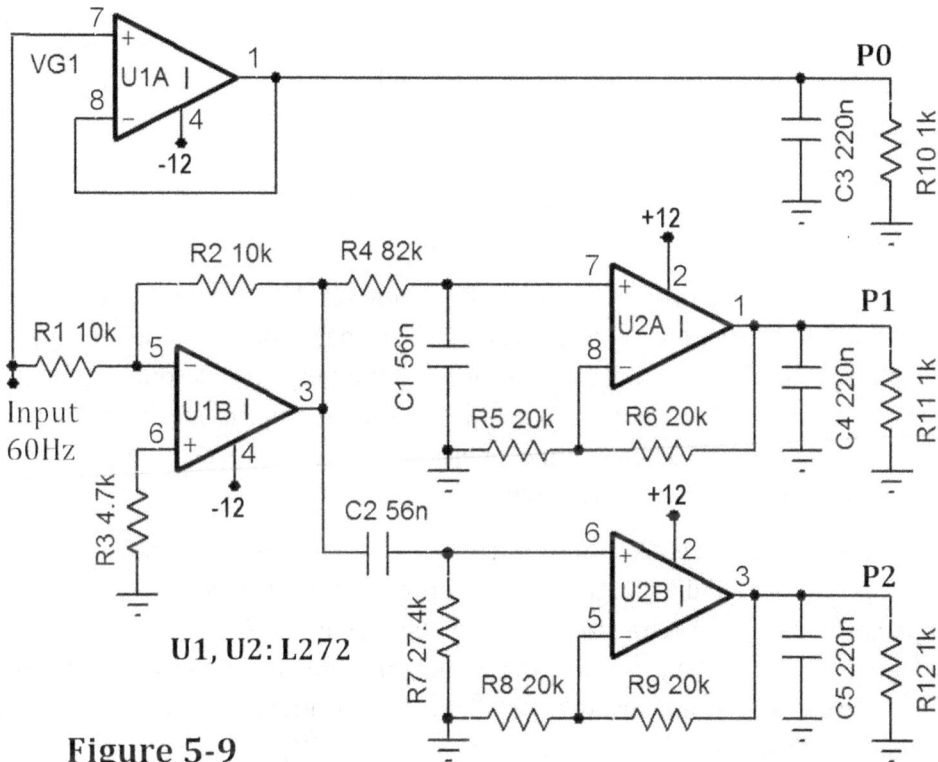

Figure 5-9

2. With the power supplies off, connect +12V to U1, pin x, and to U2 pin x. Connect -12V to U1, pin y, and to U2 pin y. Use 100µF capacitors to bypass the +12V and -12V supplies to ground.

3. Turn on the power supplies.

4. Set the function generator to produce a 12V p-p, 60.0Hz, sine wave. Connect the oscilloscope channel 1 to output P0 and channel 2 to P1. Set both channels to AC input and 2 volts per division. Set the trigger to channel 1 and the time base to 2mS per division. Center both traces. The output P0 should be exactly 12V p-p.

5. Check that the positive slope zero crossing of P1 occurs about 5.56mS before the positive slope zero crossing of P0. Refer to the diagram below.

6. Connect oscilloscope channel 2 to output P2. Check that the positive slope zero crossing of P2 occurs about 5.56mS after the positive slope zero crossing of P0. Refer to the diagram below.

61

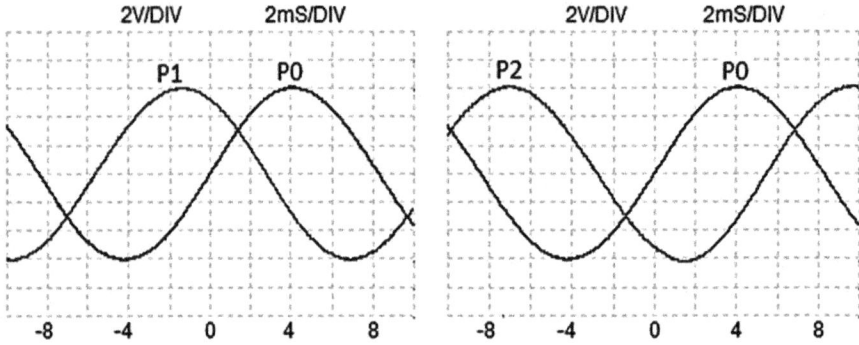

Load Limitations

Loads may be any configuration with the limitation that the power should not exceed 125mW per phase. At 12 volts peak-to-peak this is about 30mA RMS per phase. Each phase output is supplied by a very low output impedance power op-amp. Excessive current will cause the op-amp to overheat and shut down. The op-amp may resume operation when the excessive current is removed and the op-amp cools down.

The lowest impedance per phase that can be connected to the power supply depends on the load configuration. For example, at 12V p-p, the lowest impedance for a 4-wire wye would be about 140 ohms per phase (4.24VRMS/0.03A RMS). As a general rule, any impedance greater than 300 ohms per phase is safe to use with any load configuration. *The op-amps will get hot and should not be touched with loads above 100mW per phase.*

Another consideration is the power dissipation of the load components. Lab experiments using ¼ watt resistors should be designed so that the resistors don't get too hot. To avoid hot op-amps and hot ¼ watt resistors, use impedances greater than 400 ohms per phase.

Measurements

This is a wye-connected source with a common connection that is also the circuit ground for the power supply and op-amps. Non-isolated instrument grounds cannot be connected to the output of an op-amp because that would cause a short circuit. Measurements should be made with respect to the common ground, G, unless the measuring instrument is known to be isolated.

The Phase Tripler may be used as a delta source by not using the neutral, G, connection. However, measurements should be made with respect to the neutral, G, unless the instrument is isolated from the lab electrical system ground.

Chapter 6: Magnetic Applications

Magnetic fields are typically sensed by the induction effect or the Hall Effect. Induction sensors include dynamic microphones and vibration/seismic sensors. Hall sensors may be used to sense motion, position, and rotation. They are also used as magnetometers to measure magnetic field strength.

Hall Effect Sensors

A charge moving in a magnetic field is deflected by a force given by the equation: $F = qvB \sin\theta$.

B is the magnetic field flux density in Webers per square meter, q is the charge in Coulombs, v is the velocity in meters per second, F is the force in Newtons, and θ is the angle between the velocity v, and the magnetic field, B.

Electrons flow through the block on the right from left to right as indicated by the current I. The force F causes them to be deflected to the negative terminal. This creates a voltage across the block proportional to the current I and magnetic field B. This is the Hall voltage, V_H.

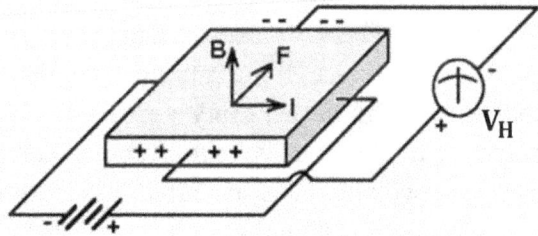

Most Hall sensors are actually semiconductor ICs. Hall Effect theory for semiconductors is similar to that for metals. The basic difference is that the motion of both the holes and electrons in the semiconductor crystal must be considered.

Induction Sensors

Induction sensors are based on Faraday's law of induction: $V_L = -N\dfrac{d\Phi_B}{dt}$. N = number of turns of wire. Φ_B is the magnetic flux through each turn. The geophone sensor on the right has a coil of wire suspended by leaf springs over a magnet. Motion of the coil relative to the magnet generates a voltage. Dynamic microphones are similar in principle.

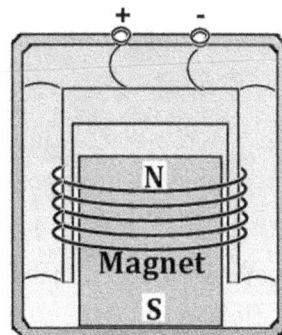

Geophone Cross-section

Project 5: Hall Effect Magnetometer

Magnetometers are used to measure and map magnetic fields. Geologists, geophysicists, and archaeologists use magnetometers for mineral and petroleum exploration, geological mapping, geophysical research, and archaeological prospecting. Instruments used in these applications are relatively sophisticated and expensive.

Hall Effect sensor ics are typically used to sense magnetic fields in the 1 Gauss to 1000 Gauss range. These sensors are commonly used in robotic applications to sense rotational speed, and angular and linear position. This project uses a Hall Effect sensor to sense and measure magnetic fields in the range of 1 to 200 Gauss. Its output may be plugged into a digital multi-meter (DMM) set to read 0 to 2 volts, where 2 volts represents 200 Gauss. Note: The earth's magnetic field has an average strength of about 0.5 Gauss.

Design Principles

An SS495 Hall Effect sensor from *Honeywell* is used in this project. It has a typical magnetic field sensitivity of about 3.125mV per Gauss. Its zero magnetic field output voltage is about one half of its supply voltage. It is available in a mini-inline pin package shown on the right.

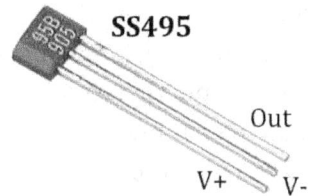

SS495

Out

V+   V-

Its output voltage varies linearly with the magnetic field strength and polarity as shown on the right.

The sensor's output voltage can be scaled with an op-amp to suit a particular application. In this project the output is scaled to produce 10mV per Gauss (1 volt per 100Gauss).

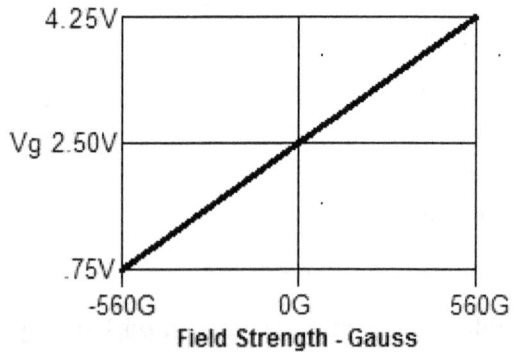

4.25V

Vg 2.50V

.75V

-560G   0G   560G

Field Strength - Gauss

The magnetometer circuit in figure 6-1 on the next page has a "zero-center" balance control to compensate for offset errors and voltage drift. It can operate from a 9 volt battery and its output can connect directly to a multimeter.

In figure 6-1 U1A is an amplifier and low pass filter. Its voltage gain is set by the ratio of R2 to R1, 39k/12k. A voltage divider consisting of R3, R4, and R6 is used to apply 2.5 volts to U1, pin3. R6 is used to set the zero Gauss offset voltage to one half of the supply voltage.

A 5 volt linear regulator is used to supply a stable voltage to the SS495 sensor. U1B is a voltage to current converter. R7 could be a zero-center, -2mA to +2mA ammeter. The DMM or voltmeter connected across R7 must be isolated from the circuit's ground.

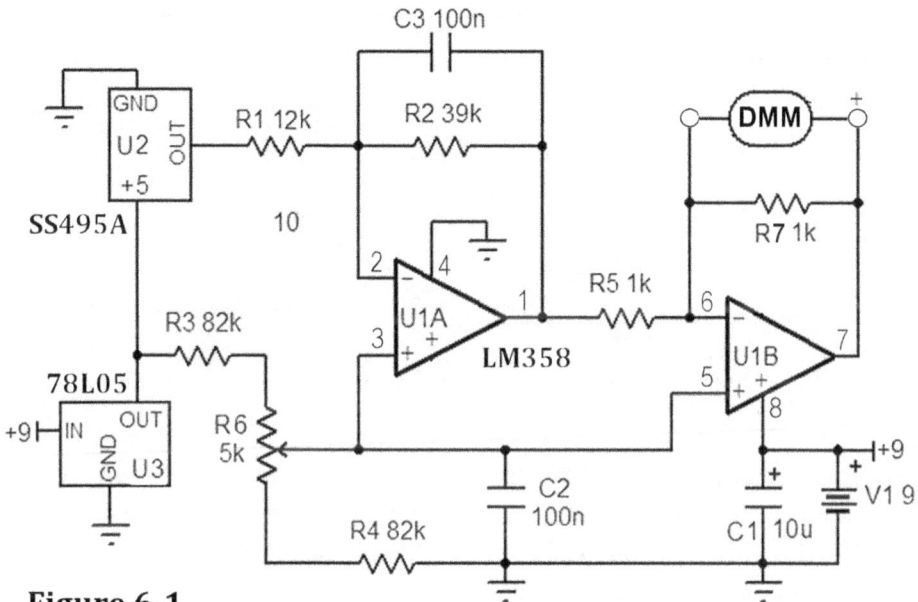

Figure 6-1

Simulation

Figure 6-2 below shows the LTspice of the magnetometer. V2 sets the positive input voltage to 2.5V on both op-amps. V1 simulates the output of the Hall sensor. The LM358 model was imported from *TI*. An LT1366 rail-to-rail op-amp could be used instead.

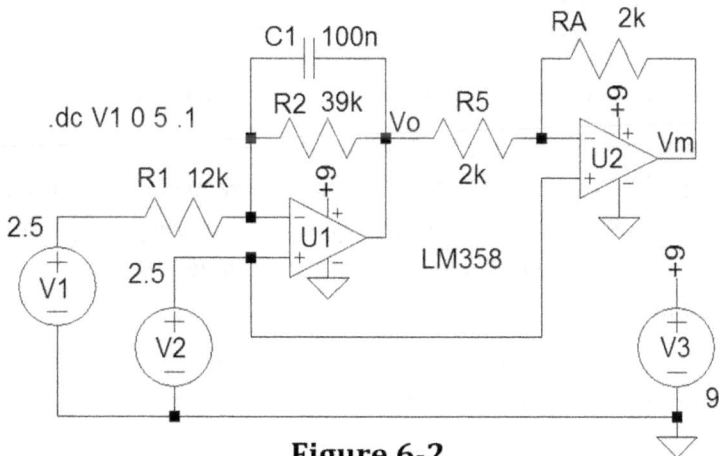

Figure 6-2

65

The graph below shows the results of a DC sweep analysis when V1 is swept from 0V to 5V. Note that the meter current is zero when the sensor output voltage is 2.5V, which corresponds to zero gauss. The direction of the meter current indicates the field polarity (North pole or South pole).

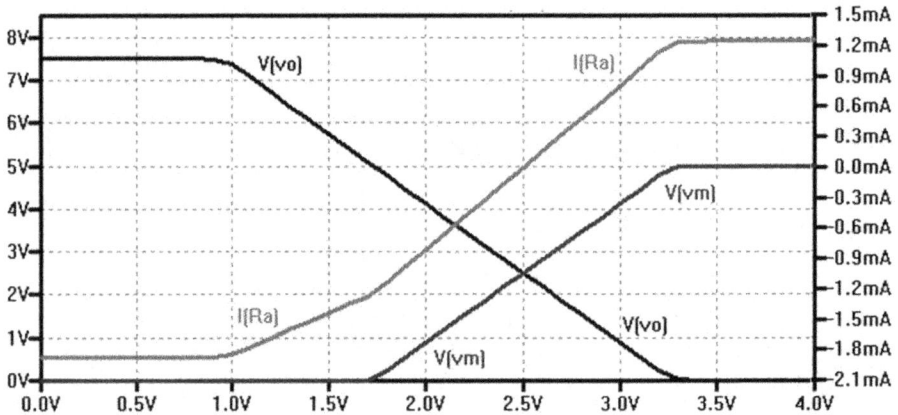

The meter current is a linear function of the sensor voltage between ±1.25mA. The voltage across the 2k meter resistor varies between ±2.5 volts. At 10mV per Gauss this corresponds to a range of ±250 Gauss.

Experiment

Parts

R1, R2: 10k. R3, R4: 82k. R5: 1k. All ¼W, 5%. R6: 5k trim pot.
C1: 10µF, 16V Electrolytic. C2, C3: 100nF.
U1: LM358 or equivalent. U2: SS495A. U3: 78L05.
9VDC battery or power supply. Meter.

Construction and testing

1. Build the circuit in figure 6-1 on a breadboard. Use a trim-pot for the 5k potentiometer. Connect a DMM across R7 to measure the output voltage.

2. Apply 9VDC to the circuit and verify that the pot will vary the output voltage. Set the output voltage to zero. The earth's field will have a small effect on the zero point. For the greatest accuracy, align the detector perpendicular to the earth's field. Use a small magnet to test the response of the circuit.

3. The sensitivity of he circuit can be changed by changing the value of R2. Set R2 to 390K to increase the sensitivity to 100mV/Gauss.

Project 6: Geophone Amplifier

This project demonstrates the application of low pass filters in a low noise low frequency amplifier circuit. The circuit was designed for a low impedance source such as a magnetic seismometer sensor or geophone. This project is about the design, simulation, building, and testing the geophone amplifier with the following specifications:

Design specifications: Input impedance: 400 ohms nominal.
Bandwidth: 0.2Hz to 140Hz.
Band-pass voltage gain: 100X.
Noise: depends on selected low noise op-amps.

Design Principles

Figure 6-3 below shows designed circuit. U1's voltage gain depends on the impedance of the sensor, G1. A 390Ω transducer would result in a gain of about 40. U2A is a unity gain 2-pole Butterworth filter. U2B is a low-pass filter and an amplifier with a gain of 2.5.

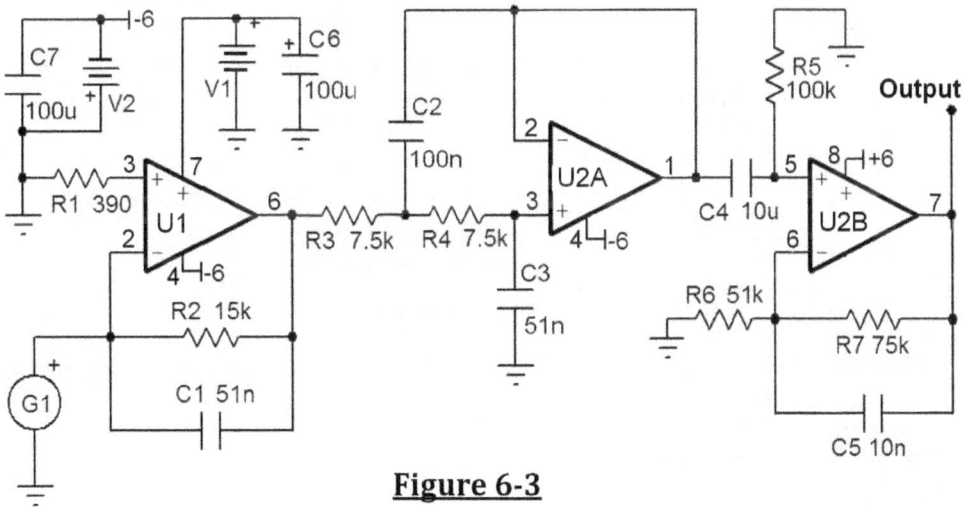

Figure 6-3

Overall this amplifier is also a 4-pole band-pass filter. The parameters were fine tuned with *LTspice* to provide a bandwidth of 140Hz. C4 and R5 make a high-pass filter with a cutoff frequency of 0.2Hz.

Offset voltages are not an issue for this circuit when precision low noise op-amps are used. An OP27 is recommended for U1 and an OP270 is recommended for U2. However, less expensive op-amps could be used in some applications.

Simulation

Figure 6-4 below shows the *LTspice* circuit diagram. Ideal op-amps are used for the simulation. Note the ".*lib opamp.sub*" directive.

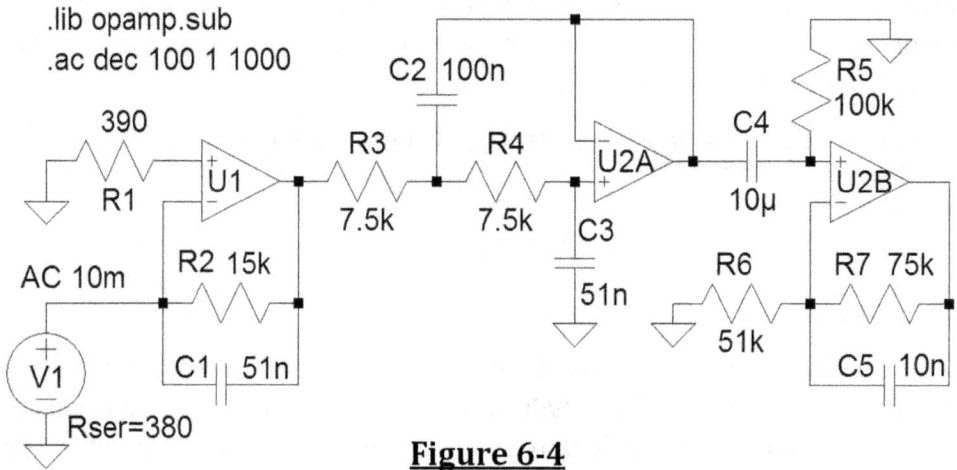

Figure 6-4

V1 simulates the magnetic sensor. Information on sensors may be obtained by searching "seismic sensors" and "geophones" on the internet. Right click on V1 to set its internal resistance (380 ohms here) and AC amplitude to 10mV. U2B is set to a gain of 2.5 with R6 equal to 51k.

The frequency response of the circuit below shows a low cutoff frequency of 0.2Hz and a high cutoff frequency of 140Hz. At 1kHz the response drops to about -42dB corresponding to an attenuation factor of over a 100.

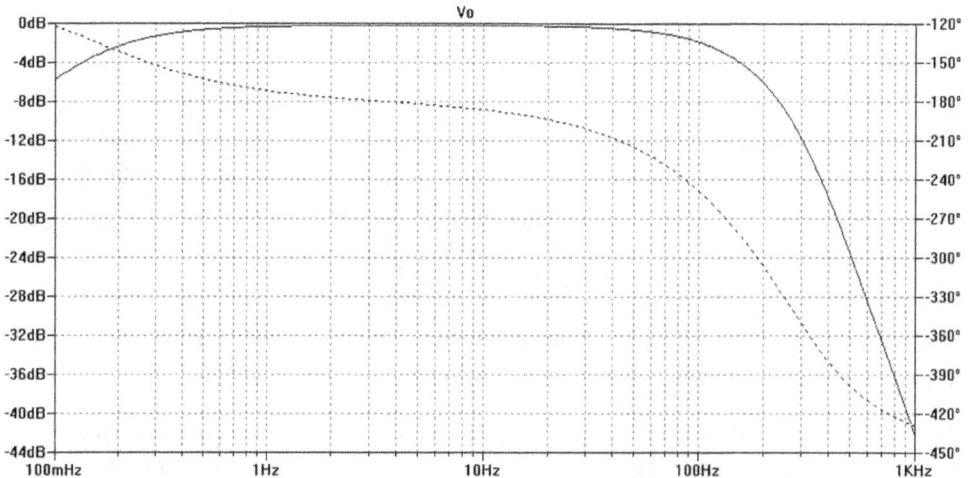

Experiment

Parts

U1: OP27. U2: OP270. Observe pin numbers.

R1: 390. R2: 15k. R3, R4: 7.5k. R5: 100k. R6: 51k/5.1k.

R7: 75k, all resistors are ¼ Watt, 5%.

C1, C3: 51nF, 5%. C2: 100nF, 5%. C5: 10nF, 5%.

C4: 10uF, 10%, non-polarized ceramic (Mouser part #810-FK20X7S1H106K).

C6, C7: 100uF, 16VDC electrolytic.

G1: Geophone, 4.5Hz, 395Ω or equivalent (R. T. Clark 4.5Hz is about $60).

Construction and testing

1. Build the circuit in figure 6-3 on a breadboard using the part values above. Wire and component leads should be kept short. See the example picture on the right.

2. Connect a function generator in series with a 390Ω resistor to the input. It will be used for G1 to test the circuit. The resistance of an RTC 4.5Hz geophone is 395 ohms. The circuit can be easily modified tor othe sensor resistances.

3. Apply power to the circuit and set the function generator to output a 40Hz, 20mV peak-to-peak, sine wave.

4. Connect an oscilloscope to the output. You should observe a 40Hz sinewave with about a 1.8 volt peak-to-peak amplitude.

5. Vary the function generator frequency to find the amplifier's corner (-3dB) frequencies where the signal amplitude is 0.707 of maximum.

6. Measure the amplifier's output at 1kHz.

7. Calculate the expected gain and bandwidth of the amplifier. Compare your results to the simulated results.

8. Calculate the worst case output noise over the bandwidth of the amplifier based on the op-amp specifications. Measurements may also show environmental noise.

The oscillograph below shows the frequency response of the amplifier for a 2 decade sweep of the input frequency. The function generator was set to do a log sweep of the frequency from 10Hz to 1kHz in 1.2 seconds. The oscilloscope was triggered by the function generator and set to sweep at 100ms per division (there are 12 horizontal divisions).

The amplifier's cutoff frequency can be found more accurately by manually sweeping the function generator to find where the amplitude is 70.7% of maximum. In this example, this turns out to be 145Hz.

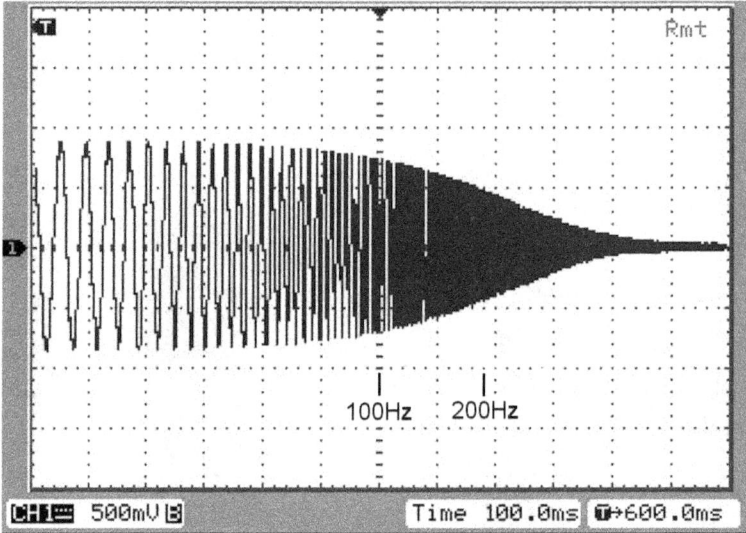

The display below shows the response for a linear sweep from 0.1Hz to 1.2kHz.

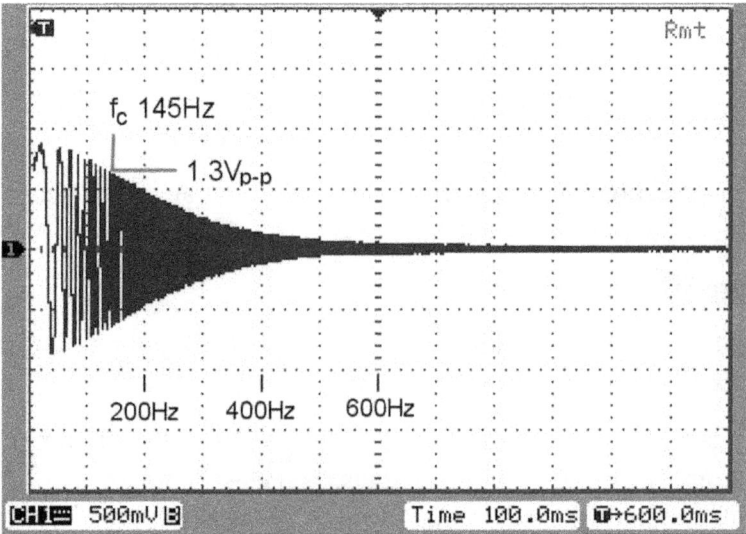

An RTC 4.5Hz 395Ω geophone was placed on the bench and the bench was tapped lightly twice. The top graph shows the time response at 100ms per division. The bottom graph shows the FFT spectrum of the response.

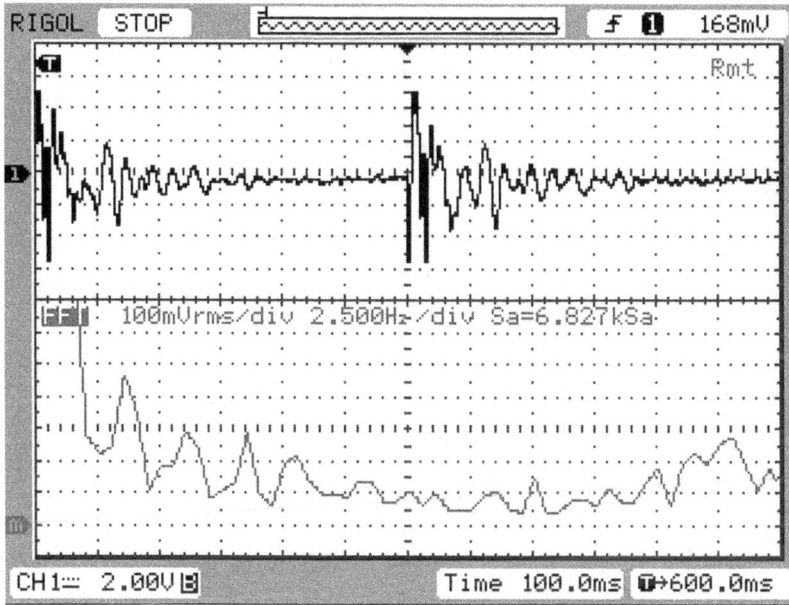

Lots of information on geophones is available on the internet. A variety of geophones are available on eBay for about $10 and up. Low frequency geophones such as the RTC 4.5Hz sell new for about $60. 1Hz geophones are also available but they are much more expensive. Geophones are generally used to sense vibrations in the 1Hz to 200Hz range, typically for geophysical exploration.

Project 7: Sensor Amplifier and Geophone damping

This project is similar to the geophone amplifier except that its bandwidth extends from DC to about 20Hz. Seismic sensors for earthquake detection are expensive. They can typically detect very low frequency (millihertz) vibrations (referred to as "long period"). Geophone frequency response is typically between 1Hz and 200Hz (referred to as "short period"). The second part of this project extends the low frequency response of the geophone for inexpensive earthquake detection.

Figure 6-5 below shows the circuit diagram of the amplifier. It is also a 4-pole low-pass filter with a cutoff frequency of 20Hz. A 60Hz notch filter is included to minimize 60Hz line interference (U2B).

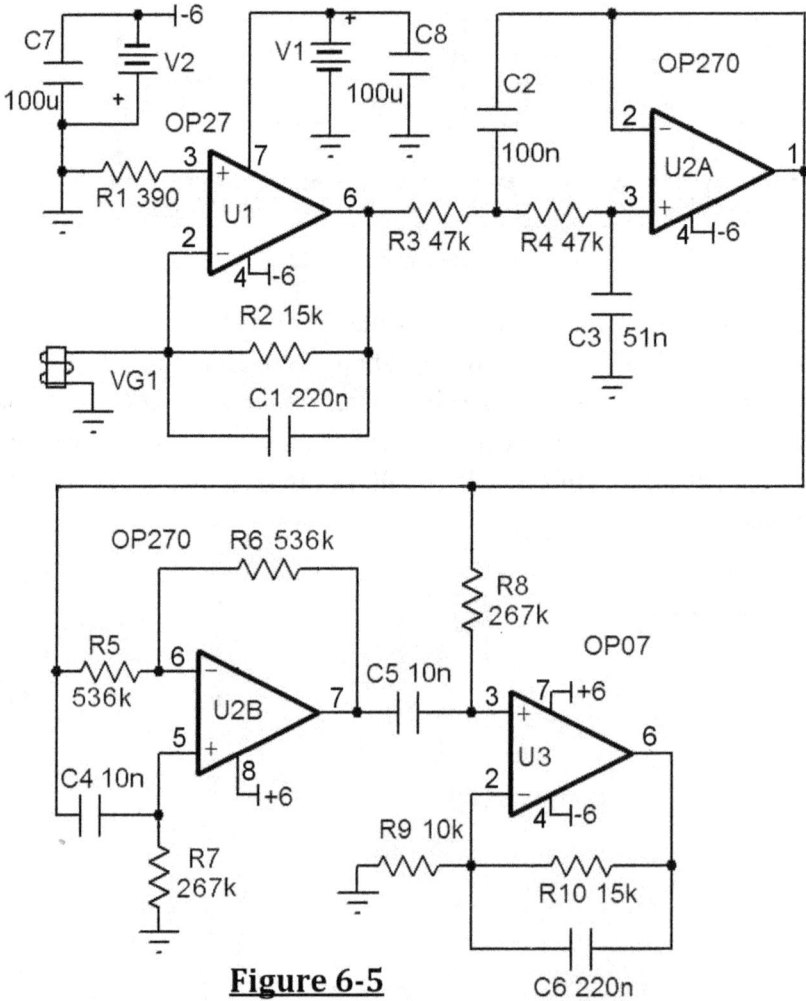

Figure 6-5

Simulation

Figure 6-6 below shows the *LTspice* circuit of the amplifier. V1's internal resistance was set to 380 ohms to represent the winding resistance of the sensor. U2A is a 2-pole low-pass Butterworth filter. U2B is a 60Hz notch filter. U3 serves as a buffer for the notch filter and provides additional voltage gain.

The notch filter's notch frequency could be changed to 50Hz for locations with 50Hz line interference but the amplifier's bandwidth would be reduced. Notch filters tend to have a relatively wide -3dB bandwidth.

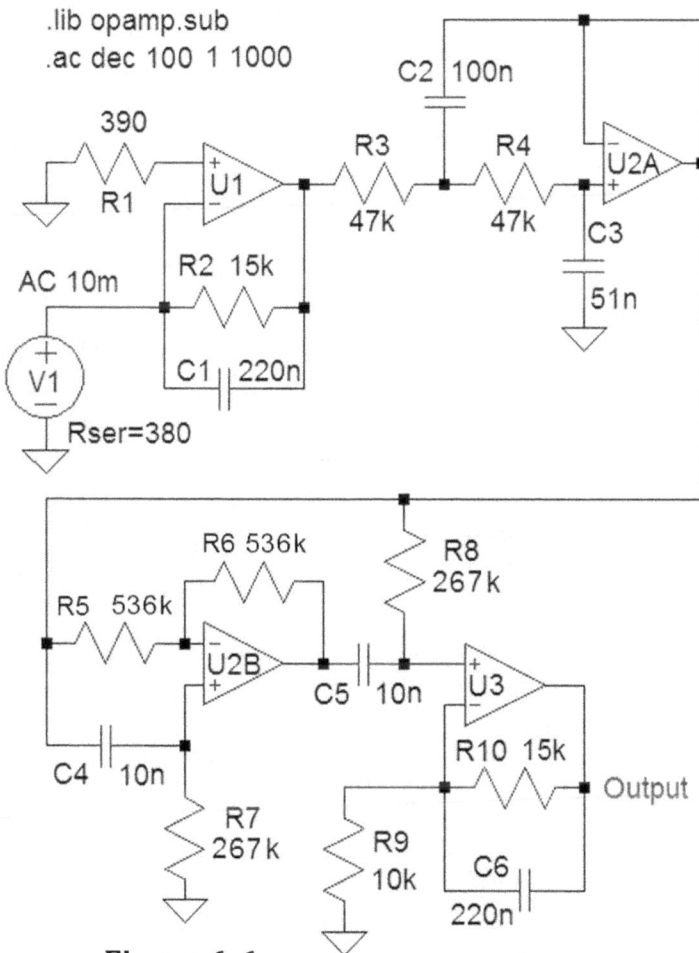

Figure 6-6

V1's AC amplitude was set to 10mV. Be sure that its internal resistance is set to 380 ohms. AC analysis was used to sweep the frequency from 1Hz to 1kHz.

The result below shows that the cutoff frequency is about 20Hz and that the attenuation at the notch frequency is 65dB.

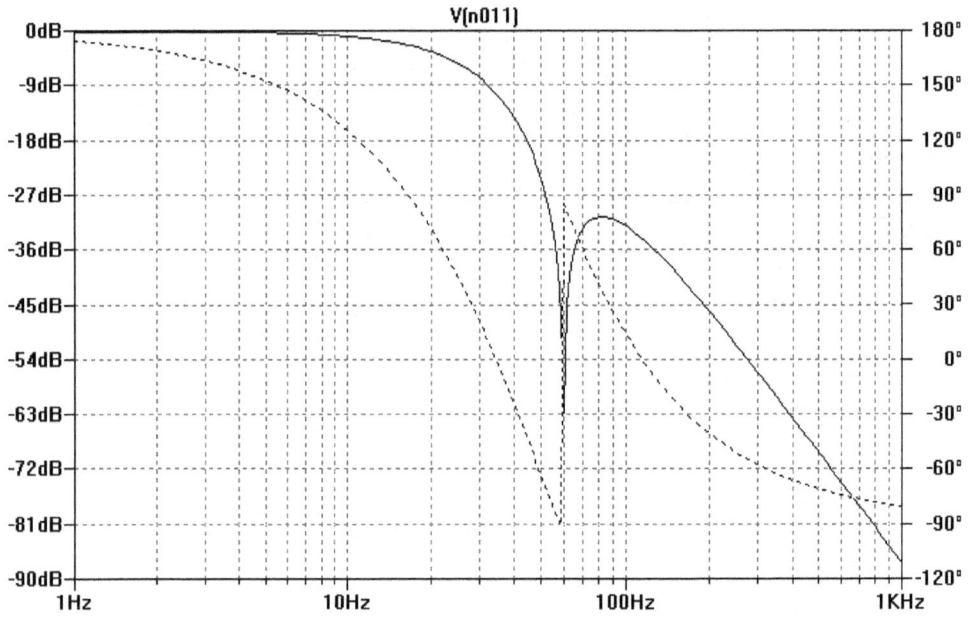
V(n011)

Change the vertical scale to linear to get the result below. The attenuation at the notch frequency and beyond appears more dramatic.

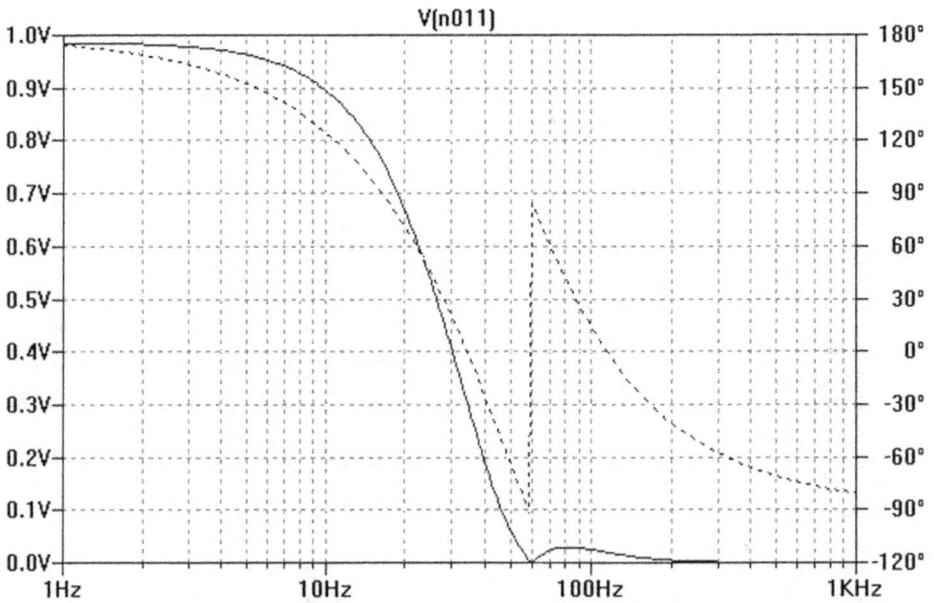
V(n011)

Experiment

Parts

U1: OP27, U2: OP270, U3: OP07.   Observe pin numbers.
R1: 390,   R2: 15k,   R3, R4: 7.5k,   R9: 10k,   R10: 15k, all ¼ watt, 5%.
R5, R6: 536k, ¼ watt, 1%.   R7, R8: 267k, ¼ watt, 1%.
C1, C6: 220nF, 5%.   C2: 100nF, 5%,   C3: 51nF, 5%.
C4, C5: 10nF, 1% (Mouser part # 80-C330C103F1G).
C7, C8: 100uF, 16VDC electrolytic.
VG1: magnetic sensor. (Function generator with 330 series resistor for testing).

Construction and testing

1. Build the circuit in figure 6-5 on a breadboard using the part values given above. Wire and component leads should be kept short.

2. Connect a function generator in series with a 390Ω resistor to the input. It will be used for G1 to test and analyze the circuit.

3. Apply power to the circuit and set the function generator to output a 40Hz, 30mV peak-to-peak, sine wave.

4. Connect an oscilloscope to the output. You should observe a 40Hz   sine wave with an amplitude of about 3 volts peak-to-peak.

5. Vary the function generator frequency to find the amplifier's corner frequencies (-3dB – where the signal amplitude is 0.707 of maximum).

6. Vary the function generator frequency to find the amplifier's notch frequency and notch attenuation.

7. Calculate the expected gain and bandwidth of the amplifier. Compare your results to the simulated results.

8. Calculate the worst case output noise over the bandwidth of the amplifier based on the op-amp specifications.   Measurements may also show environmental noise.

9. Measure the amplifier's DC output voltage (offset voltage). Calculate the worst case output offset voltage of the amplifier based on the op-amp specifications.

10. U1 and U3 have offset compensation pins. Look up the required offset compensation circuits in the op-amp's data sheet.   Apply offset compensation to U1.

The oscillograph below shows amplifier response for a 2 decade sweep of the input frequency. The function generator was set to sweep the frequency from 1Hz to 100Hz in 1.2 seconds. The oscilloscope was triggered by the function generator and set to sweep at 100ms per division (there are 12 horizontal divisions). The 60Hz notch is apparent on the right side of the display.

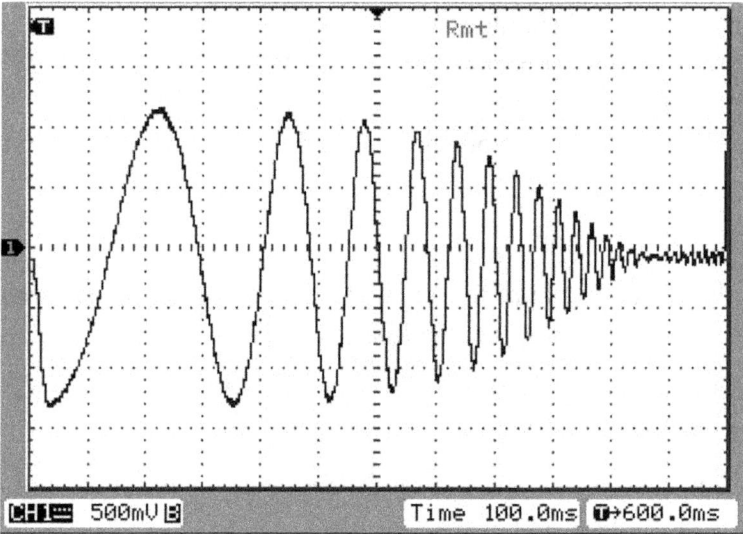

The display below shows the response for a linear frequency sweep from 0.1Hz to 120Hz. The 60Hz notch is at the center of the screen.

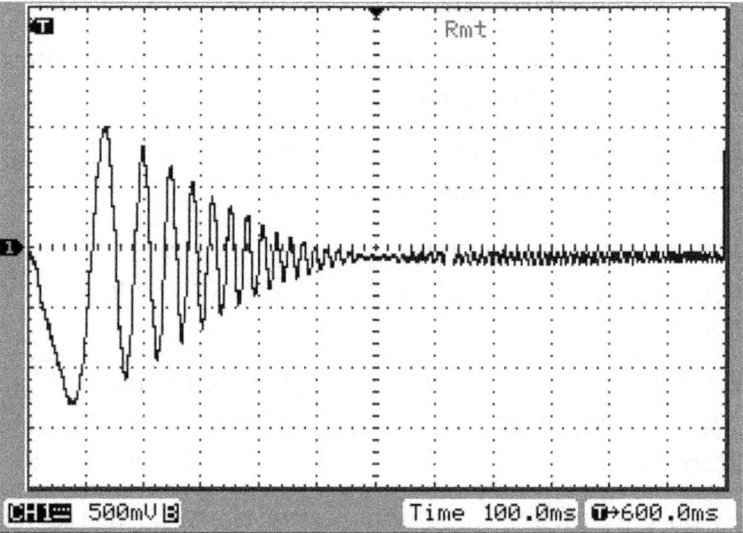

The oscilloscope's vertical sensitivity is changed to 50mV per division in the display below. Note that there is a -110mV DC offset. Input offset compensation should be applied to U1. If this amplifier's gain is increased to 1000, its output offset voltage would become -1,1V.

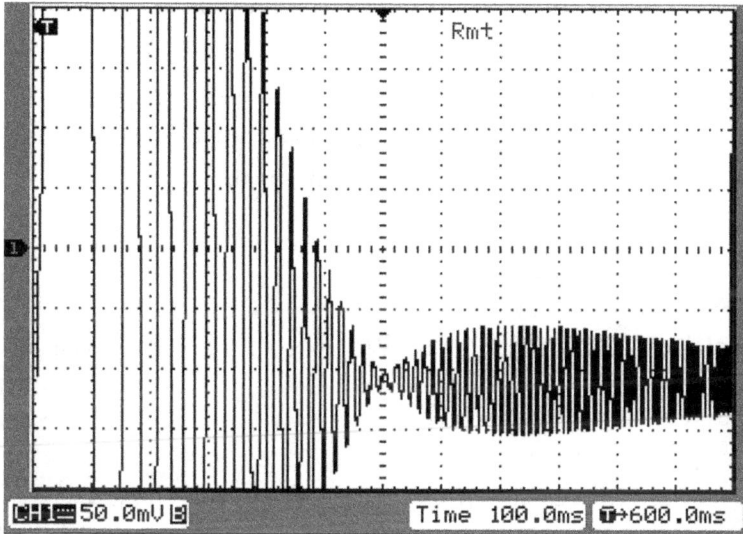

Extending the Geophone's Low-frequency Response

There is considerable interest in extending the low frequency response of geophones for earthquake detection because geophones are simpler and much less expensive than seismograph instruments[1]. Over-damping is a common approach which can be done by connecting a resistance to the geophone. However, this requires a negative resistance. According to Ulman[2] this typically requires a negative resistance: $R_d = -0.8R_c$. R_c is the coil resistance odf the geophone.

Refer to "VNIC – Negative Impedance Converter" on page 52 of this book. The impedance for the circuit of figure 3-17 is given below. A suggested circuit to replace the U1 circuit in figure 5-14 is given on the next page.

$$Zin = \frac{Vin}{Iin} = -\frac{R1}{R2}Zx.$$

1. Novel Tools for Research and Education in Seismology, by Mikhail E. Boulaenko. Master of Science Thesis, Institute of Earth Physics, University of Bergen, December 2002

2. Over-damping geophones using negative impedances, Bernd Ulman, 2005

Example

This example is intended to show a possible design approach for a VNIC with a voltage amplification of 40 and given that a 4.5Hz, 395Ω, geophone requires an input impedance of - 316Ω (Zin = -.8R$_d$ = -.8(395) = -316Ω. The calculation for a real application would require more information than just the geophone coil resistance. Specifications for the RTC-4.5-395 are given n the appendix.

The objective is to design an amplifier with a gain of 40 with a negative resistance at the inverting input of about -316Ω. Begin with the equations for the input current and the voltage on the input terminals.

$$Iin = \frac{Vin - Vo}{Rx} = \frac{V1 - Vin}{Ri} \quad \text{and} \quad Vin = \frac{R_1}{R_1 + R_2}Vo.$$

$$VinRi - VoRi = V1Rx - VinRx \implies Vin(Ri + Rx) = V1Rx + VoRi.$$

$$\frac{R_1}{R_1 + R_2}Vo(Ri + Rx) - VoRi = V1Rx.$$

$$\frac{Vo}{V1} = \frac{Rx}{\frac{R_1}{R_1 + R_2}(Ri + Rx) - Ri}, \quad \text{If } R2 >> R1 \text{ and } Rx >> Ri \text{ then } \frac{Vo}{V1} \approx \frac{Rx}{-Zin - Ri}.$$

Design: $R_d = 395$, $Zin = -0.8R_d = -316$.

For a gain of 40: $\dfrac{Vo}{V1} \approx \dfrac{Rx}{-Zin - Ri} = -40 \approx \dfrac{Rx}{316 - 400} \implies Rx = 40(84) = 3360.$

let R2 = 10k, calculate R1:

$$Zin = -\frac{R1}{R2}Rx = -\frac{R1}{10000}3360 = -316 \implies R1 = \frac{3160000}{3360} = 940.$$

The input impedance and voltage gain is sensitive to the values of R1, R2, and Rx. Therefore standard value 1% resistors are chosen for the design.

The circuit in figure 6-7 was fine tuned with *LTspice*. Rx = 3240. R1 = 953, R2=10000 (all 1%).

.lib opamp.sub
.dc V1 -.05 .05 1m

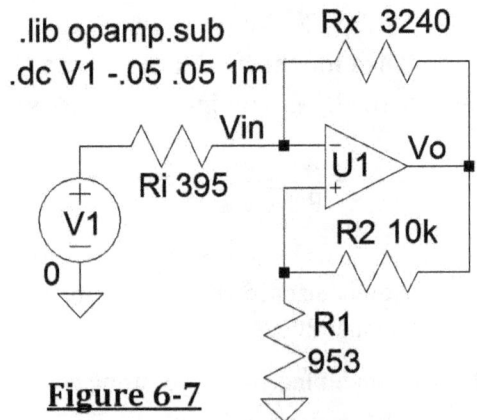

Figure 6-7

Simulation Results

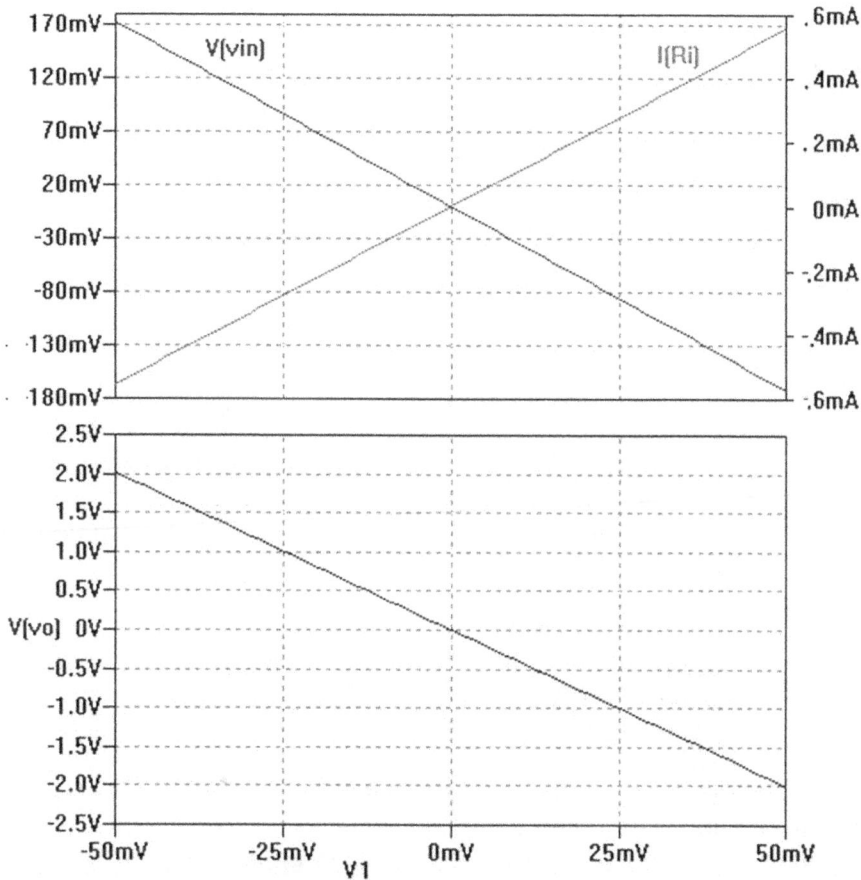

The top graph shows the input voltage at the negative input of the op-amp and the current flowing into it. The total change in the input voltage is -340mV. The total change in the input current is 1.12mA. This shows an input resistance of -304Ω.

The bottom graph shows that the total change in the output voltage is 4 volts. The corresponding change in the input voltage, V1, is 100mV. This corresponds to a voltage gain of 40.

This VNIC could be the first amplifier, U1, of the geophone circuit figure 5-14. It would connect directly to the Butterworth filter. Theoretically it should extend the low frequency response of the RTC 4.5 to about 0.5 Hz. However, the negative impedance here was calculated based only on the sensor's coil resistance. There are other factors, such as the sensor's coil mass, that need to be considered to calculate a more accurate value of negative resistance.

Chapter 7: Low Frequency Receivers

Op-amp circuits may be used in the design of radio frequency receivers. Low "voltage noise" op-amps are generally used for low impedance sources (un-tuned coil antennas) while low "current noise" FET op-amps are generally used for high impedance sources (un-tuned wire or stylus antennas). Op-amp filters are used as needed for a particular application.

The ELF "Extremely Low Frequency" band (3Hz to 30Hz) is used to study "natural" radio emissions due to natural phenomenon such as lightning, volcanic eruptions, and Ionospheric disturbances. Also studied is the effect of the "earth-ionosphere" wave guide (Schumann resonance) whose resonant frequency is about 8Hz (186,000 miles per second/24,000 miles = 7.75Hz).

The ULF "Ultra Low Frequency" band (300Hz to 3kHz) and VLF "very low frequency" band (3kHz to 30kHz) are also used to study "natural" radio emissions due to natural phenomenon such as lightning, volcanic eruptions, and Ionospheric disturbances. In particular, the frequency range of 20Hz to 20kHz is used to listen to natural radio emissions because these frequencies are in the audio range. This frequency range can be directly input to the microphone input of a PC and processed by software. More information is available on the internet. A popular program for the PC can be downloaded at: http://www.qsl.net/dl4yhf/spectra1.html

The VLF "Ultra Low Frequency" band range from 9kHz to 30kHz is used for submarine communication because of the water propagation capability of VLF frequencies. Tuned VLF receivers are also used to monitor the received signal strength of VLF stations to detect ionization phenomenon in the earth's upper atmosphere such as that caused by solar flares. Information is available on VLF solar flare receivers and SID "Sudden Ionospheric Disturbance" receivers on the internet.

The loop antenna can replace the geophone in chapter 5, projects 6 and 7. Project 6 would become an amplifier for radio frequencies in the super low frequency band (SLF) while project 7 would become an amplifier for the extremely low frequency band (ELF).

Using a wire antenna for the ELF band (3Hz to 30Hz) is simpler and less expensive than using a loop antenna. However, an efficient antenna would need to be several miles long at these frequencies. Short antennas may be used but they present a very high impedance (over 100 meg-ohms). A very high input impedance amplifier with extremely low current noise is required.

Project 8: ELF Receiver Design

LTspice is a very effective tool for designing low frequency op-amp receiver circuits. Different op-amps can be tested in the same circuit to evaluate their performance. Many op-amp models are available in LTspice plus manufacturers often provide models of their op-amps that can be imported into LTspice. The noise performance of the first amplifier stage is especially important. This project is an exercise in LTspice design and performance comparison.

This LTspice project uses the same circuit as project 7, except the low input impedance U1 circuit is replaced with a high input impedance circuit. It is designed to detect the electric field component of the extremely low frequency radiation. An FET input op-amp is used because the FET input stage has very low bias currents and very low current noise.

An efficient antenna for very low frequencies would need to be many miles long. This amplifier is intended to be used with a short wire antenna (under 8 feet) so that the circuit responds to the electric field strength at the antenna.

The amplifier circuit in figure 7-1 replaces the U1 circuit in project 7. Its output connects directly to the Low-pass filter, U2A. It has an input impedance of about 20 meg-ohms and a voltage gain of 45 in the pass-band. The LT1055 op-amp is recommended as one of the lowest current noise, wide bandwidth, op-amps available.

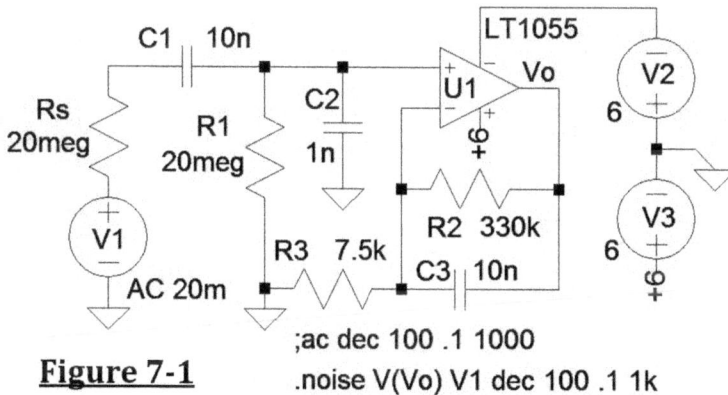

Figure 7-1

;ac dec 100 .1 1000
.noise V(Vo) V1 dec 100 .1 1k

Type of receiver is very susceptible to environmental noise, especially from power lines, and to micro-phonics effects. ELF receivers are usually operated far from cities and power lines. The performance of a particular design can be evaluated by simulation. The performance of a prototype will also be affected by the circuit layout, environmental noise, and even mechanical considerations.

Experiment: LTspice Design Evaluation

Below: Simulation of the amplifier's frequency response. The 20mV source is divided by Rs and R1 so that 10mV is applied to the op-amp's positive input. Output is about 450mV in the pass-band (dBV = 20*log*(0.45) = -6.94dBV).

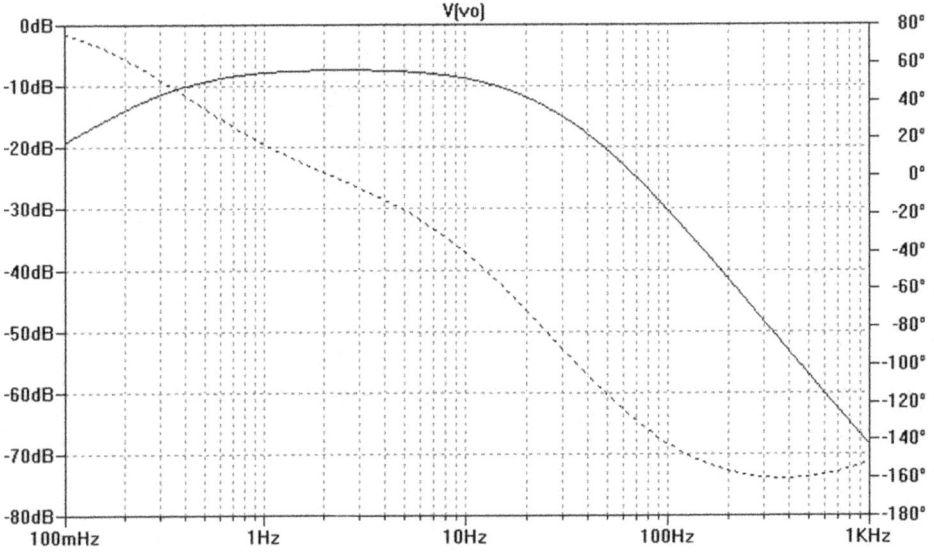

Below: Simulation result for the noise response of the amplifier. Average output noise in the pass-band is about $18\mu V/\sqrt{Hz}$. About $12\mu V/\sqrt{Hz}$ of that is due to the amplifier's input resistance.

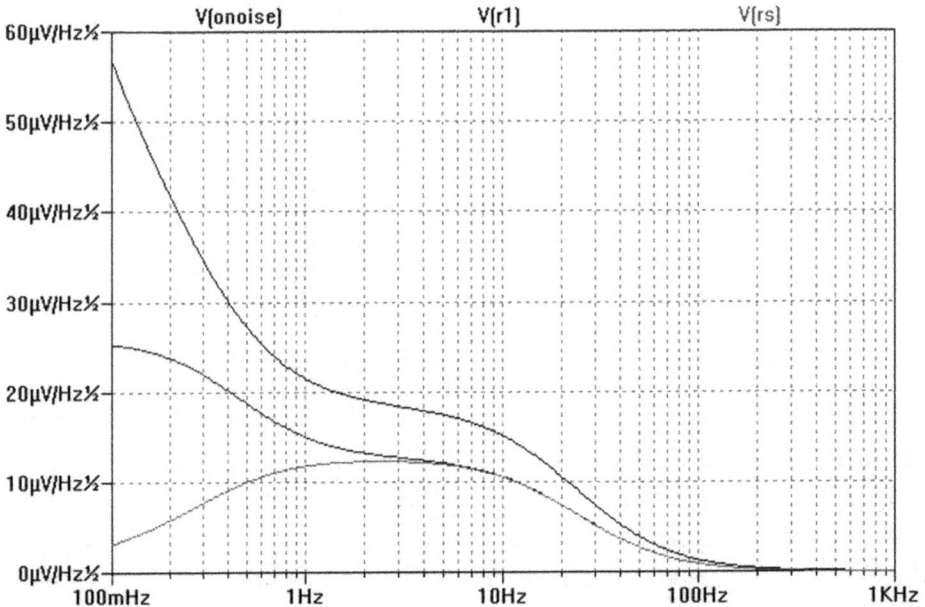

Exercises

1. Replace the LT1055 with other FET input op-amps, such as the LT1112. Compare the frequency response performance and noise performance of the two op-amps. Download the datasheets for the two op-amps [try pdf downloads from Digi-Key]. Compare their bandwidth and noise data.

2. Replace the LT1055 with a bi-polar op-amp, such as the OP27. Compare the frequency response performance and noise performance of the two op-amps. Download the datasheets for the two op-amps [try pdf downloads from Digi-Key]. Compare their bandwidth and noise data.

3. Perform an AC analysis on the complete ELF receiver circuit in figure 7-2 below using ideal op-amps. Set simulation to decade sweep, 100 points per decade, 1Hz to 1kHz.

4. Replace the op-amps with your selection of models; such as a LT1055 for U1 and OP-27s for the rest. Do an AC analysis. Set simulation to decade sweep, 100 points per decade, from 1Hz to 1kHz. Perform a noise analysis and explore the noise contributions of the op- amps and resistors.

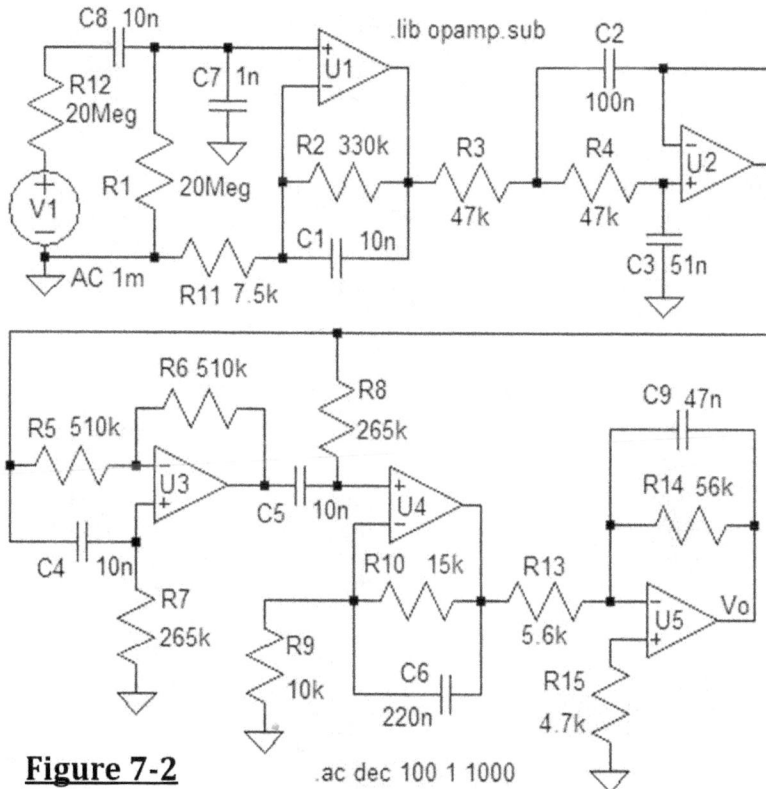

Figure 7-2

Project 9: Broadband VLF Receiver Design

The objective of this LTspice project is to design a high input impedance broadband VLF receiver for the "audio" frequency range of 300Hz to 10kHz. This is primarily an LTspice design and analysis exercise. It involves selecting op-amps based on gain, bandwidth, slew rate, and noise characteristics. This receiver is very susceptible to environmental noise from power lines, trucks, automobiles, and electrical devices.

Design requirements:
Bandwidth: 300Hz to 10kHz   Band-pass voltage gain: 300.
Input impedance: greater than 20 Meg-ohms. Noise: As low as practical.
60Hz rejection: greater than 72dB.
Power requirement: 9V battery.

Figure 7-3 represents a possible design. A low "current noise" op-amp is used in the input stage. Low-pass and high-pass Butterworth filters are used for a band-pass filter. A 60Hz notch filter, U2B, is buffered by U3A. Filters, U1B, U2A, and U3A, have a unity band-pass gain. The total gain of U1A and U3B is about 300. U1A and U3B also provide additional low-pass filtering.

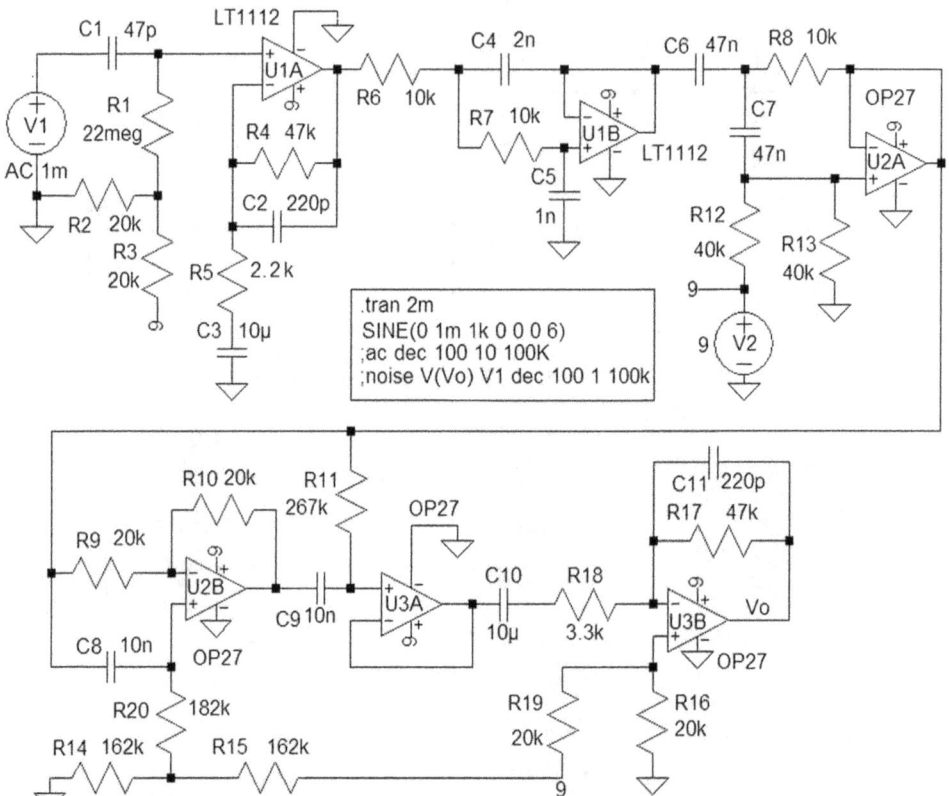

Figure 7-3

84

Experiment: LTspice Receiver Analysis

Price and performance of op-amps varies considerably. A simulation program such as *LTspice* allows the designer to compare the performance of a variety of op-amps in the same circuit. Manufacturers often supply spice models for their op-amps which can be imported into LTspice. Information on importing third party models may be found on *Linear's* web site and by a web search. More information is also presented in project 12.

Electronics supply companies such as *Digi-Key* and *Mouser* provide links to manufacturer's data sheets and possibly to their spice models. Part cost, availability, and specifications can easily be compared. Also, an equivalent or similar *LTspice* library part may be found to use in a simulation.

This receiver design uses the same band-pass filter circuit as project 7, except for different cutoff frequencies. The 60Hz notch filter provides additional 30dB suppression at the low frequency end.

AC analysis is performed from 10Hz to 100kHz with a decade sweep. Simulation results are shown below. These results were obtained by fine tuning key component values and re-simulating the circuit until a satisfactory result was obtained.

The first op-amp is the most critical, requiring low noise and wide bandwidth. A much less expensive TL072 FET input op-amp could also be used for U1.

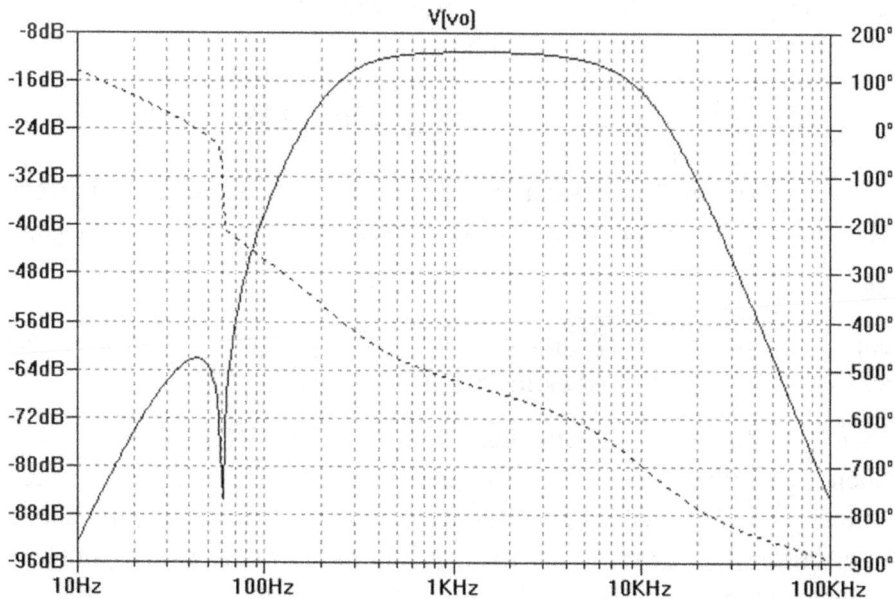

The choice of U1A will have a large effect on the circuit's noise response. A graph of the noise at the output of the receiver, V(onoise) is shown below.

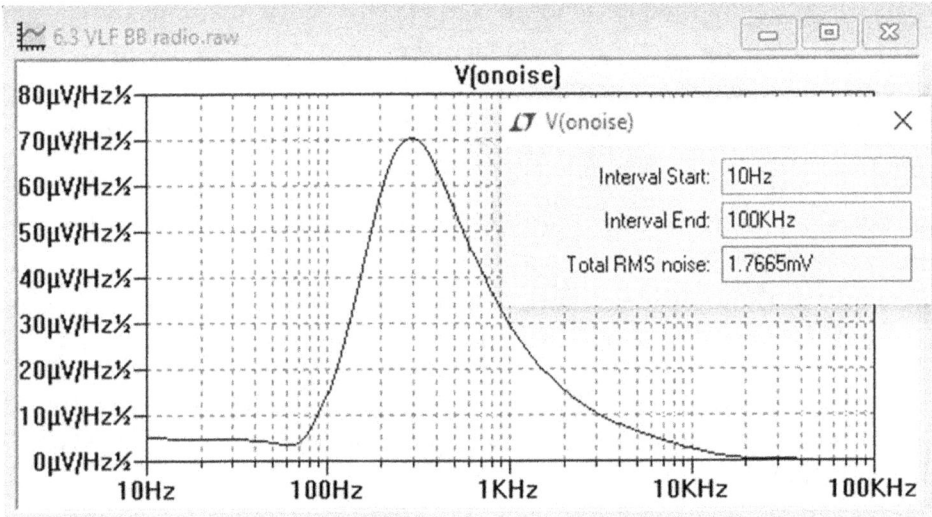

LTspice can integrate the noise over the bandwidth. Ctrl-click the data trace label, V(noise), in the waveform viewer (above). The total RMS noise for the bandwidth specified in the .noise directive is displayed. The total noise at the output of this receiver design is about 1.77mV.

Signal to Noise Ratio

The signal output over the receiver's bandwidth is about 300mV. The receiver's signal to noise ratio is calculated as:

$$S/N = \frac{300mV}{\sqrt{2}(1.77mV)} = 120 \quad (S/N)_{dB} = 20log120 = 41.6dB.$$

This is the signal to noise ratio for a 1mV peak input signal. Since the noise is relatively constant, the signal to noise ratio increases with an increase in signal.

Exercise

1. Change U1B to an OP-27 and compare the frequency response, noise response, and signal to noise ratios to when U1B was an LT1112.

2. Change U1A to an OP-27 and keep the OP27 for U1B. Compare the frequency response, noise response, and signal to noise ratios to when U1B was an LT1112. Step 1 above should prove to be the better configuration. The signal to noise ratio could be further improved, but due to the environmental noise level at these frequencies would probably not warrant the effort.

Project 10: Tunable VLF Receiver

The tuning range of this receiver is from about 15kHz to about 30kHz, depending on the particular antenna coil and variable capacitor used. It is designed to receive submarine communications stations, navigation transmitters, and time signals.

VLF frequencies can propagate in salt water to a depth of over 50 feet. High power transmitters (200KW – 1000KW) and very large antennas are used on land to send messages at a slow data rate (encrypted) to submarines.

One application of a receiver such as this is the detection of solar flares and gamma ray bursts. Solar flares and gamma ray bursts produce sudden changes in the ionization of the ionosphere. These changes effect the propagation of radio signals that are received by refraction from the ionosphere. Receivers for this application are specifically called "SID" (Sudden Ionospheric Disturbance) monitors. They are tuned to a distant VLF station and monitor and record the station's signal strength. Sudden changes in strength are interpreted as a sudden ionospheric disturbance event.

Receivers are typically tuned using inductors and capacitors, op-amp band-pass filters, simulated inductor (gyrator), and digital filters. This project uses a high-Q ferrite core antenna tuned by a variable capacitor at the input of the first amplifier, U1. Its tuning range is from about 15kHz to about 31kHz using the components specified.

An 80mH ferrite rod antenna and a 3-gang variable capacitor with a capacitance range from about 150pF to 1150pF is used in this design. Adding 150pF for stray capacitance, the tuning range of this combination is calculated below.

$$f_{min} = \frac{1}{2\pi\sqrt{.08(1.3\times10^{-9})}} = 15.6\text{kHz}, \quad f_{max} = \frac{1}{2\pi\sqrt{.08(.3\times10^{-9})}} = 32.5\text{kHz}.$$

The range can be changed with additional parallel capacitance. In addition to the tunable amplifier a band-pass filter is added to reduce out of band interference.

The total maximum voltage gain from antenna to output is about 20,000 (86dB). A gain switch is included to reduce the maximum gain to 2000. A variable gain control can further reduce the gain to about 60. Three outputs are provided: *direct* and *am* for oscilloscope observation and *signal strength* for a recorder or data logger.

The circuit is presented in figure 7-4 below. An inexpensive low "current noise" FET op-amp, the TL071, was chosen for U1 in the input stage. A wide bandwidth bi-polar input op-amp, the OP270, was chosen for U2, the band-pass filter variable gain amplifier circuit.

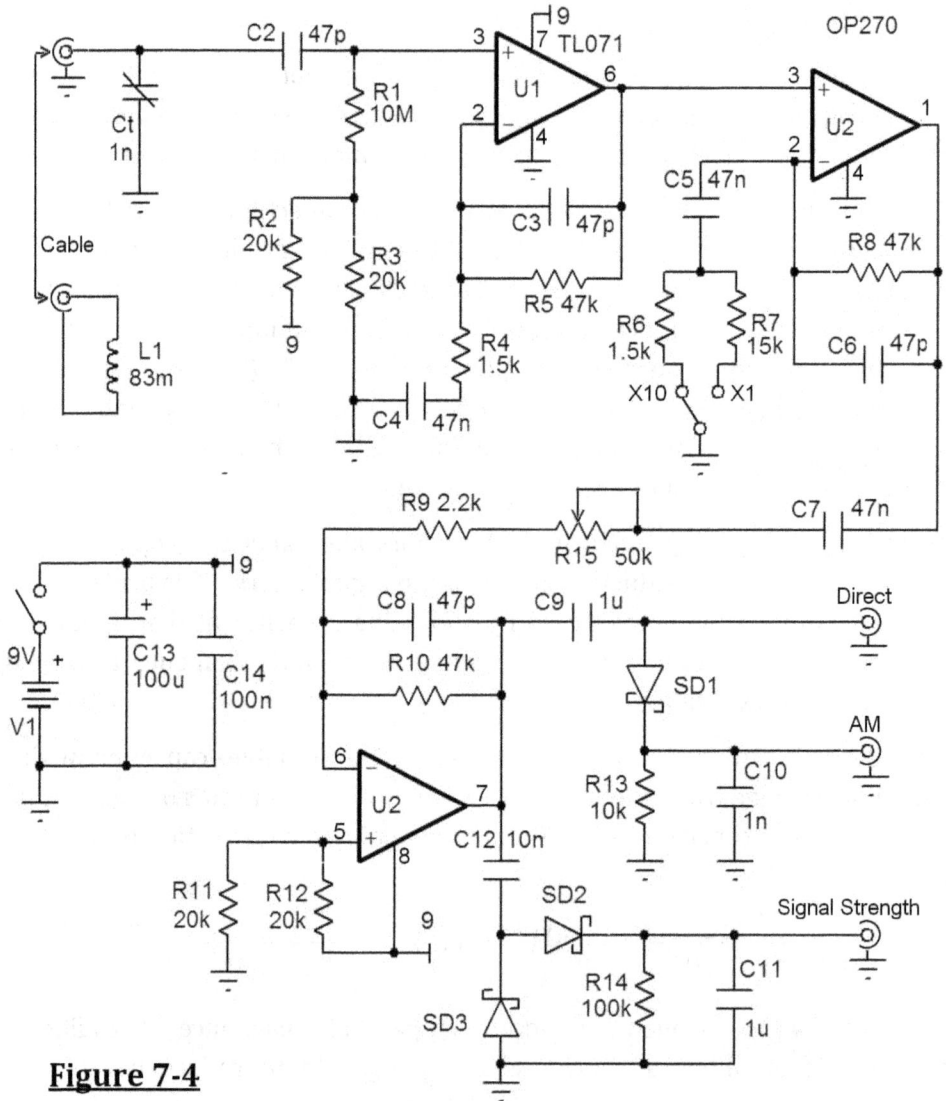

Figure 7-4

The ferrite rod antenna connects to the circuit's input through a bnc connector, J1. This is intended for a short cable, less than about one foot. Longer cable will add too much capacitance and is not suitable for this application. The antenna can also be wired directly to the input.

Simulation

The simulation below shows the frequency response of the receiver without the tuning circuit. The signal is applied directly to the BNC antenna input.

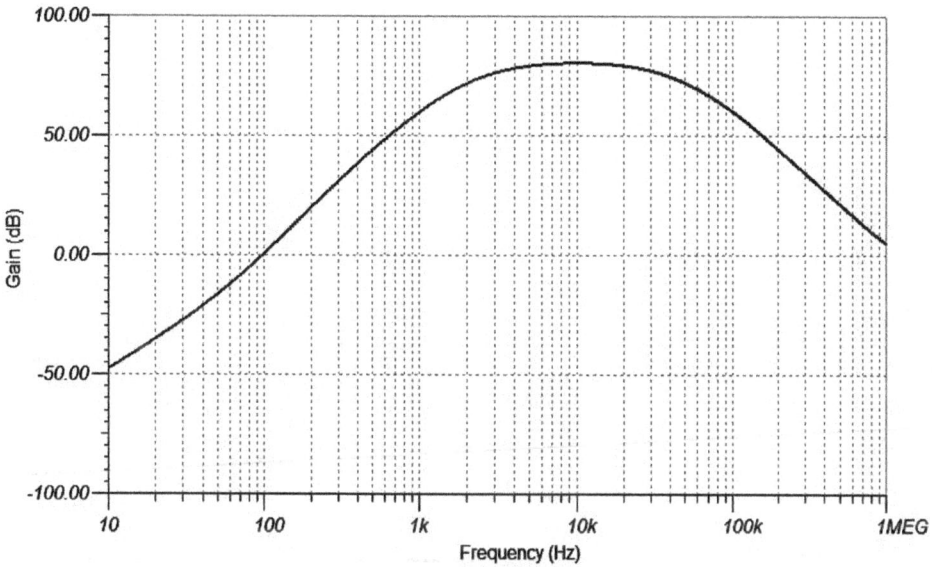

The simulation below shows the response of the first amplifier when tuned to 17.3kHz.

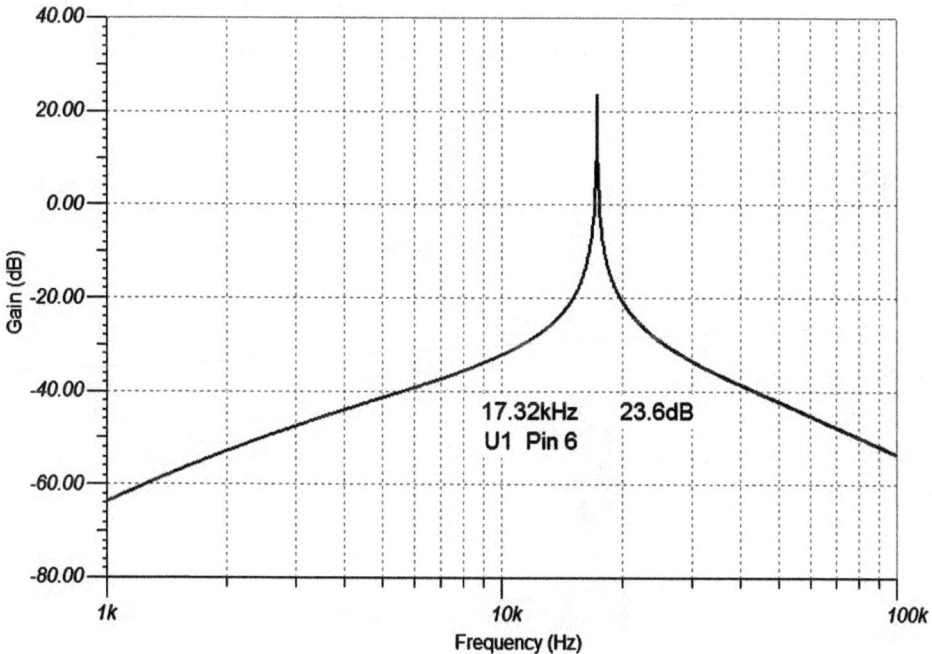

The simulation below shows the noise response at the output of the first stage when tuned to 17.3kHz. A 17.3kHz, 1mV peak amplitude sinusoid, is applied to the input through a 10 megohm resistor. Most of the noise is due to the 10 megohm input resistors.

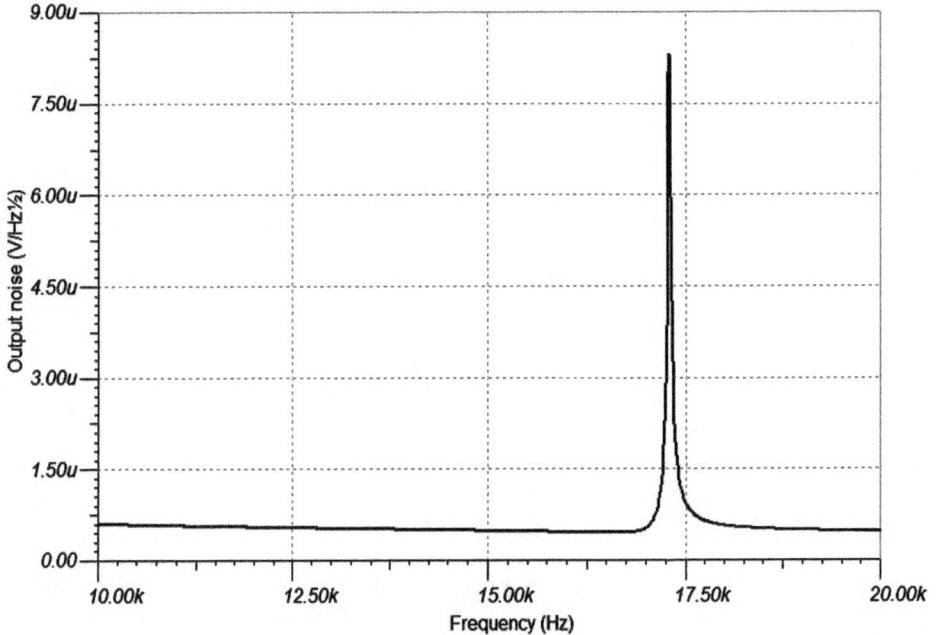

This simulation was performed by *TINA* (Toolkit for Interactive Network Analysis). A free version is available from *Texas Instruments* (www.ti.com).

Experiment

A picture of the receiver project is shown on the right. The 3 sections of the tuning capacitor are connected in parallel to give a capacitance range of about 150pF to 1150pF. The tuning range of this reciever is 15kHz to 32kHz.

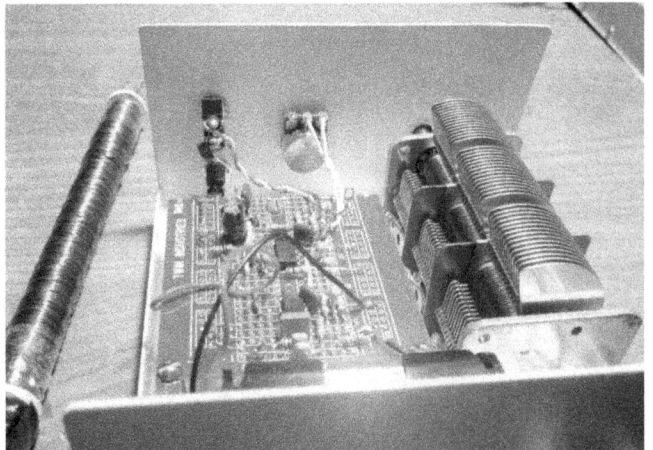

The 80mH ferrite rod antenna is shown on the left side of the receiver.

A frequency sweep is performed using a function generator and oscilloscope. The function generator is set to sweep from 20kHz to 32kHz at a rate of 120ms per sweep. The oscilloscope is synched to the function generator with a sweep rate of 10ms per division. The Rigol oscilloscope has 12 horizontal divisions therefore each horizontal division represents 1kHz.

The function generator is coupled to the receiver by placing a wire near the ferrite rod antenna. The receiver is tuned to 26kHz and adjusted for an output amplitude of about 3V$_{p-p}$.   The oscilloscope display below was obtained by setting the display to "dots" and "infinite persistance".

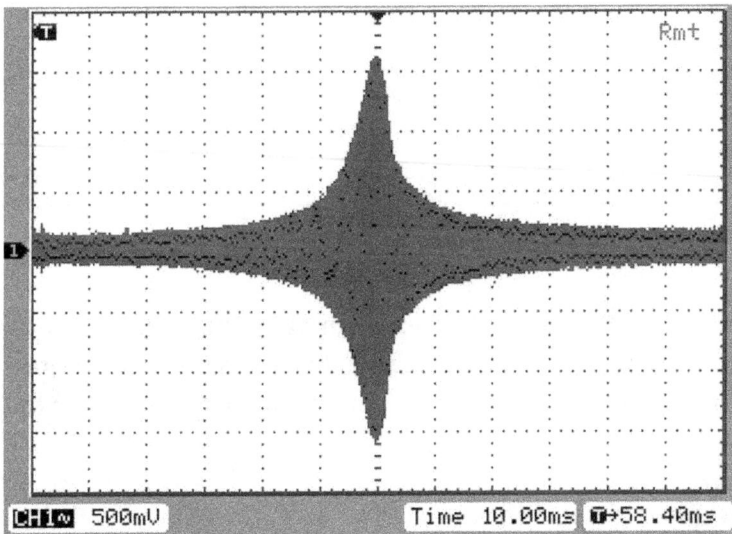

Above: The -3dB bandwidth is about 500Hz corresponding to a Q of 52.
Below: Simulation of the receiver using ideal components shows a -3dB bandwidth of about 60Hz.

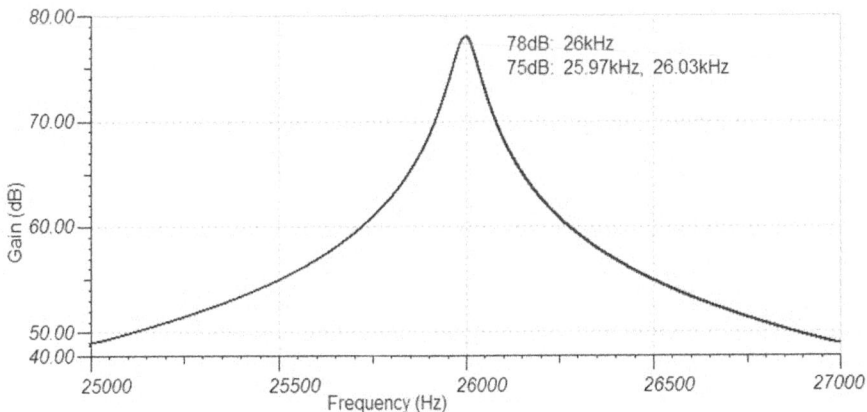

The Q of the 80mH inductor was measured to be about 57 which would result in a 450Hz bandwidth at 26kHz. This is equivalent to connecting a 200 ohm resistor in series with the inductor as shown by the simulation results below.

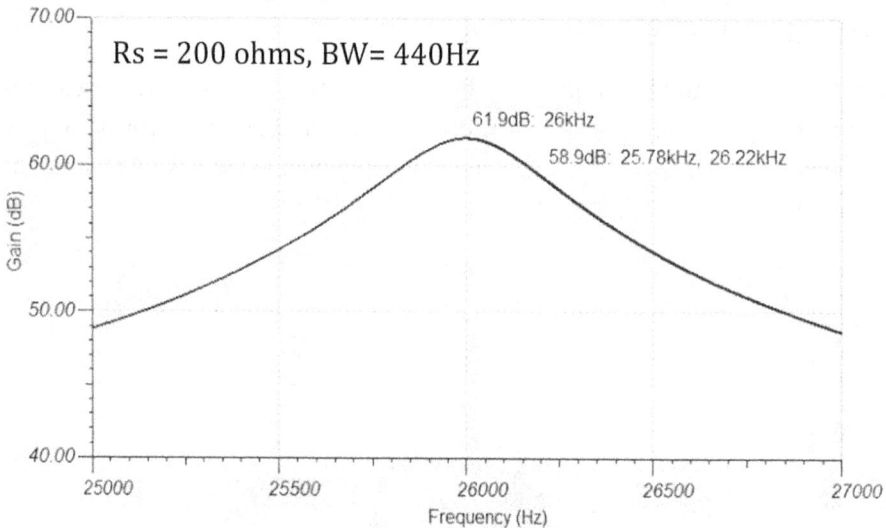

Rs = 200 ohms, BW= 440Hz

61.9dB: 26kHz

58.9dB: 25.78kHz, 26.22kHz

The inductor's series resistance is less than one ohm so that it has a negligible effect on its Q. Core losses in the ferrite rod antenna have the greatest effect on the inductor's Q. Other losses in the reciever's input circuit reduce the Q to 52.

Operation

This receiver easily was able to receive the Jim Creek Naval Radio Station, NLK, at 24.8kHz, indoors with just the ferrite rod antenna (about 400 miles north of the receiver location). Jim Creek Naval Radio Station is a very low frequency, 1.2 megawatt, radio transmitter facility at Jim Creek near Arlington, Washington, north of Seattle. It sends messages one-way to submarines of the Pacific fleet while submerged. Messages may be transmitted by either on-off keying or frequency-shift keying.

VLF and SID receivers are used by scientists and hobbyists. Lots of information may be found on the internet on antenna and circuit design as well as on observatories and software.

Project 11: Gyrator Tuned VLF Receiver

The gyrator circuit is an alternative to using a variable capacitor for tuning a VLF receiver. This seems to be a popular circuit. It is considerably noisier than the capacitor tuned circuit but also considerably less expensive. It is presented here as an interesting circuit to experiment with. A low noise amplifier in front of this circuit could improve the signal to noise ratio of a receiver using it.

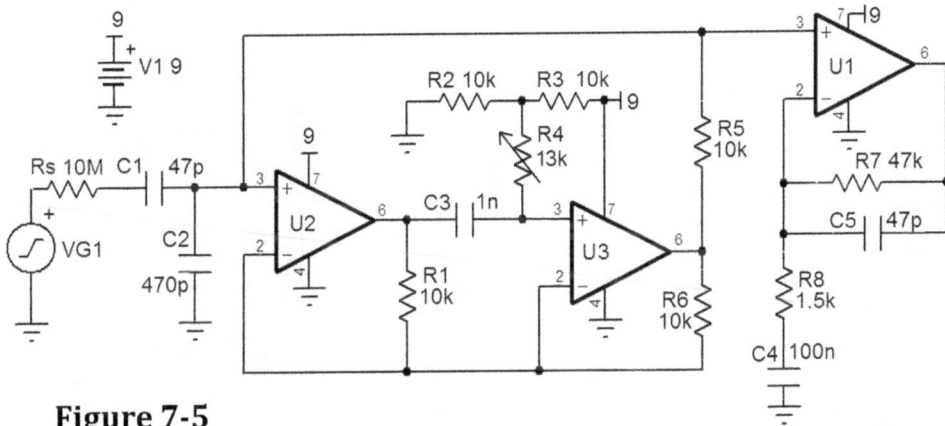

Figure 7-5

For comparison, the same op-amps and frequencies are used in this simulation as for project 10. U1, U2, and U3 are all TL071s. The frequency response below shows about the same selectivity as in project 10.

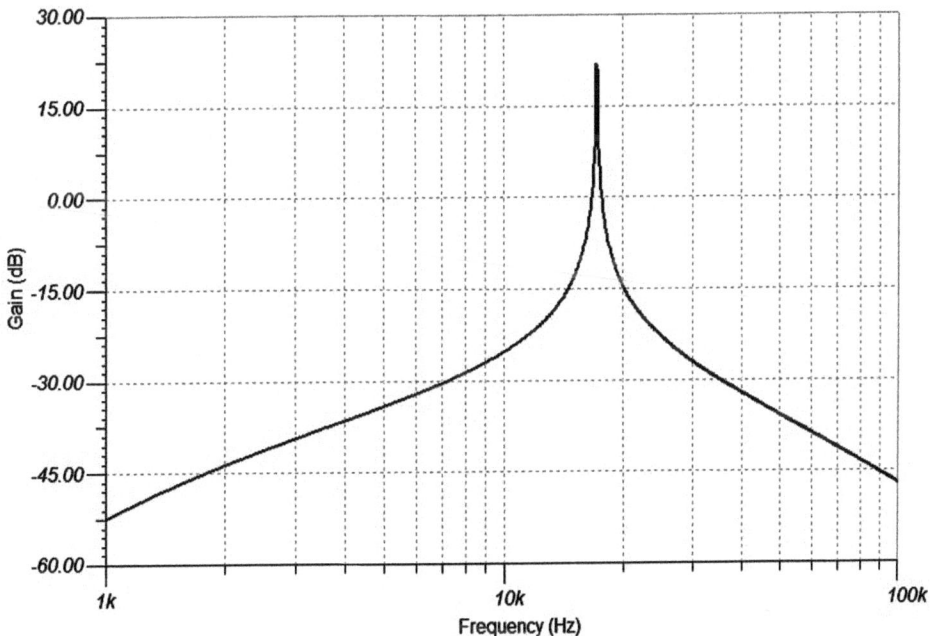

Below is the noise simulation result:

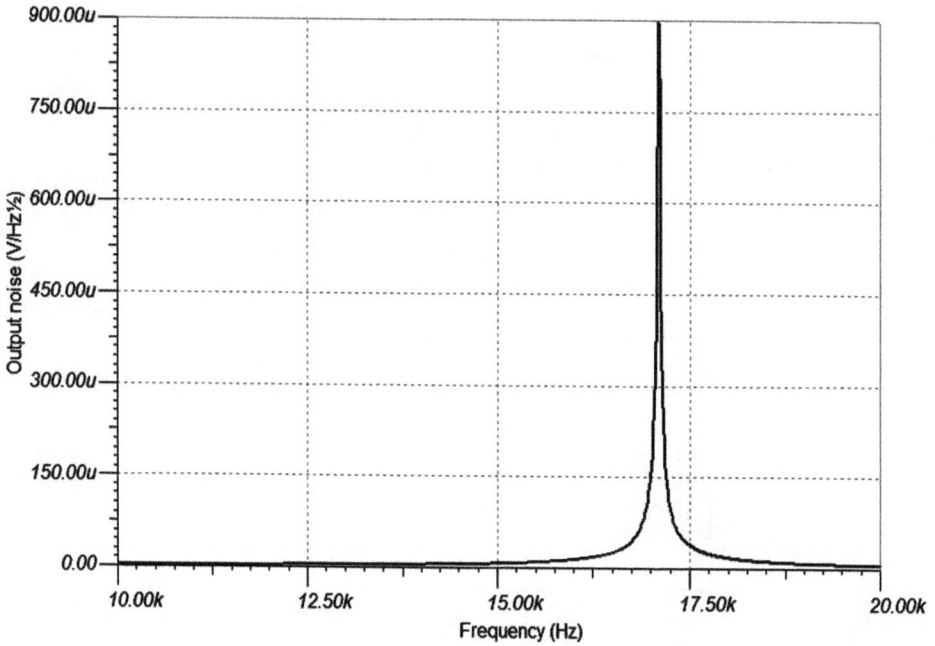

This simulation shows over a hundred times more noise voltage than the capacitor tuned equivalent in project 10. This may be reduced by selecting other op-amps. A broadband low noise pre-amplifier could be used to improve the receiver's signal to noise ratio.

Chapter 8: Photo-Resistor-Coupler Applications

Most opto-couplers use an LED and a photo-transistor or photo-diode to transfer signals between electrically isolated circuits. Photo-resistor opto-couplers use an LED and a photo-resistor. The value of the photo-resistance can be varied by the LED current.

Opto-couplers can be used in an "on-off" mode as an isolated switch or remote controlled switch. In this mode they can also be used to transfer digital data. Photo-resistor opto-couplers are commonly used as variable resistors in audio and music applications in volume control, compressor, limiter, and tremolo circuits.

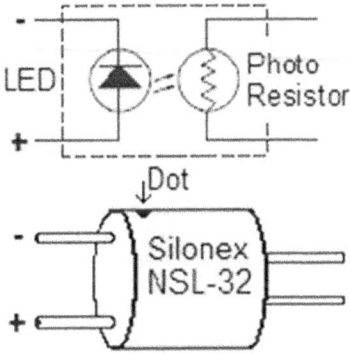

The diagram of the NSL-32 is shown above and its typical response is shown on the right.

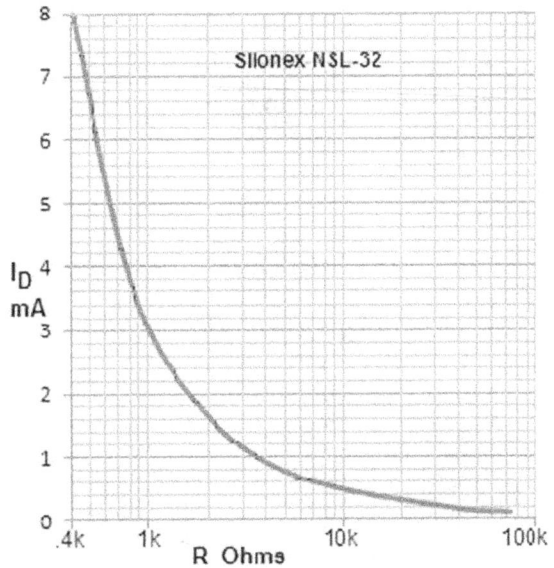

One major limitation of the photo-resistor coupler is its slow response time, on the order of tens and hundreds of milliseconds. An improved version of the NSL-32, the NSL-32SR3 has a somewhat better response time, a higher off resistance and lower on resistance. Please refer to the appendix for more information.

From the graph above one sees that the resistance is highly non-linear and that greatest range of resistance change occurs at LED currents below 1mA. A voltage controlled current source is typically used to vary the LED current.

The Silonix NSL-32 series is now produced by *Advanced Photonix, Inc* and is available from *Digi-Key* and *Newark*.

95

Project 12: Photo-Resistor Opto-Coupler Simulation

Manufacturers often provide Spice models for their products but no models appear to be available for Photo-Resistor Opto-Couplers. The simple model presented in figure 8-1 is used in the next three projects to demonstrate the basic operation of the opto-coupler. The model, named ZX1, has characteristics similar to a *Silonix* NSL-32.

Current sources I1 and I2 are used to test the model and are not part of the model. H1 and H2 are current controlled voltage sources. H1 and VS between nodes 3 and 0 simulate a voltage controlled resistor. H2 senses the diode's current and outputs the control voltage for H1. Thus the diode current controls the resistance between nodes 3 and 0. Voltage sources V1 and VS are set to zero volts and used only to sense the currents through them.

Rise and fall times are approximated with R4, R2, C1, and current control switch, W1. W1 is controlled by the direction of the current in H2.

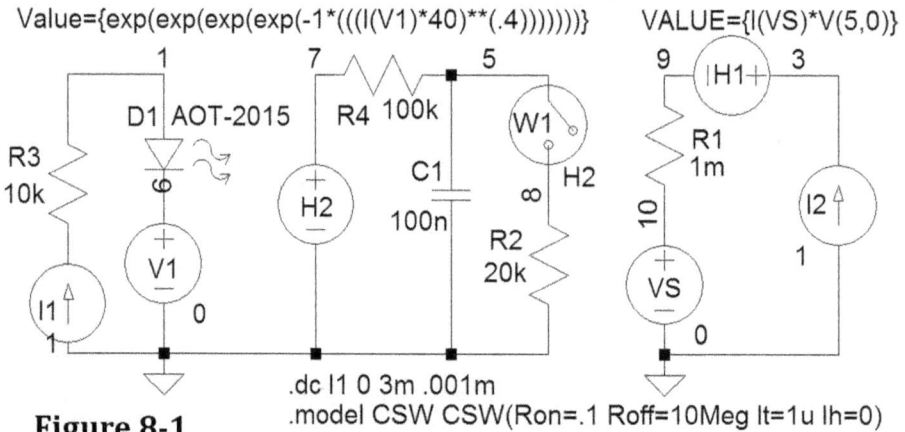

Figure 8-1

The result of the simulation of figure 8-1 is on the right. The vertical axis is the voltage drop across the simulated resistance by the 1 ampere current source so that the vertical axis represents resistances in ohms.

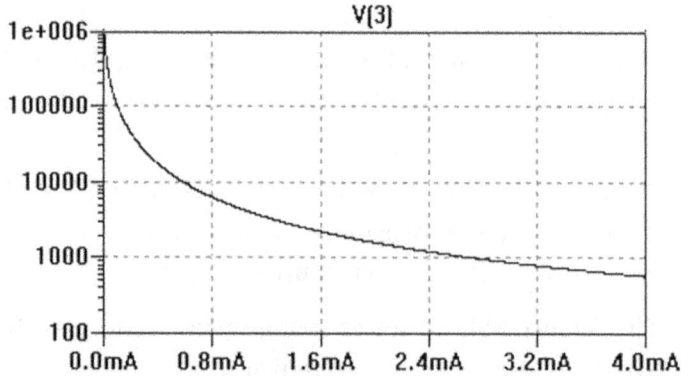

A ".subckt ZX1" file is developed next from this simulation.

96

Four exponentiations are used to obtain a non-linear response that is similar to the response of the NSL-32 photo-coupler. In the file below, the sources I1 and I2 are removed. Node 0 on the input side is relabeled node 2 and node 0 on the output side is relabeled node 4.

This file may be copied, saved as "ZX1.sub", and copied into *LTspice's* sub circuit directory (\LTC\LTspiceIV\lib\sub). A 4 pin symbol file, "ZX1.asy" needs to be copied into the "Optos" directory (\LTC\LTspiceIV\lib\sym).

```
********* + - R  R
.subckt ZX1  1 2 3 4
VS 10 4 0
R1 9 10 1m
R2 8 2 20k
H2 7 2 Value={exp(exp(exp(exp(-1*(((I(V1)*40)**(.4)))))))}
V1 6 2 0
D1 1 6 AOT-2015
H1 3 9 VALUE={I(VS)*V(5,0)}
R4 5 7 100k
C1 5 2 100n
W1 5 8 H2 CSW
.model D D
.lib C:\Program Files (x86)\LTC\LTspiceIV\lib\cmp\standard.dio
.model CSW CSW(Ron=.1 Roff=10Meg It=1u Ih=0)
.ends.model  D  D
.lib C:\Program Files (x86)\LTC\LTspiceIV\lib\cmp\standard.dio
.ends
*Note: the .lib paths may need to be changed for some pcs.
```

The table below compares three photo-coupler characteristics to the spice model derived here, which is designated as ZX1. It can be seen that the characteristics vary widely between devices. These characteristics can also vary by more than 50% for any particular device, especially for low currents.

Photo-Coupler Comparison				
Resistance	ZX1/ZX1A	NSL-32	NSL-32SR3	VTL5C2
Ω	µA	µA	µA	µA
1MEG	6	?	2	?
100K	105	100	6	200
10k	600	500	30	600
1k	2800	3000	200	3000
100	*	*	4000	*

Project 13: Voltage Divider

The circuit in figure 8-2 demonstrates an application of the resistor opto-coupler as a voltage divider (volume control).

Both couplers are driven by the same current source. Their operation as the top resistor Ut, and the bottom resistor Ub, of the divider is compared.

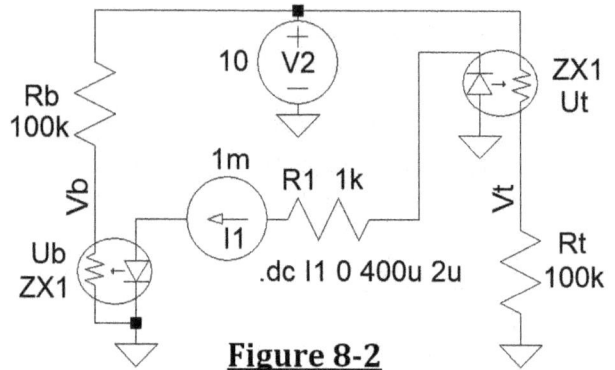

Figure 8-2

The simulation result on the right shows the two curves intersecting at 5V, where the coupler resistor values are 100k ohms. This is the basic idea for a balance control or cross-fader.

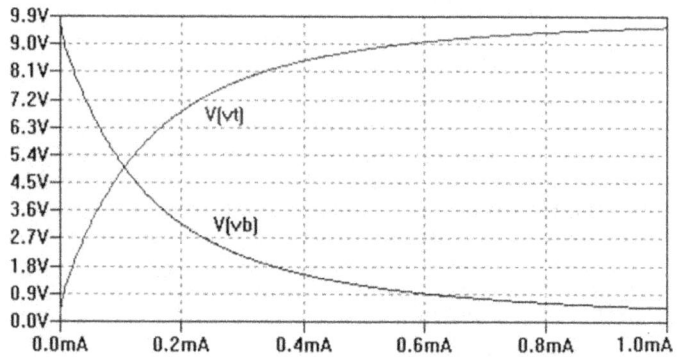

The graph below shows a 200ms transient analysis response of the ZX1. A 90mS, 1mA, current pulse is applied to The ZX1 (.1mA to 1mA). The time it takes to go from 100k ohms at 0.1mA to 4.8k ohms at 1mA is about 6ms. The time it takes to go from 4.8k ohms at 1mA to 100k ohms at 0.1mA is about 180ms.

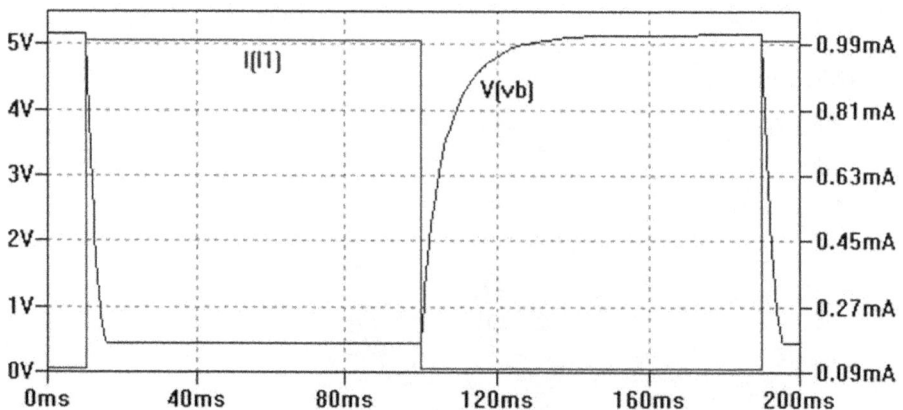

Project 14: Voltage Controlled Oscillator

Voltage controlled oscillators are possible applications for the photo-resistor output opto-coupler. Figure 8-3 below combines the voltage controlled current source circuit with a Schmitt trigger "relaxation oscillator" circuit.

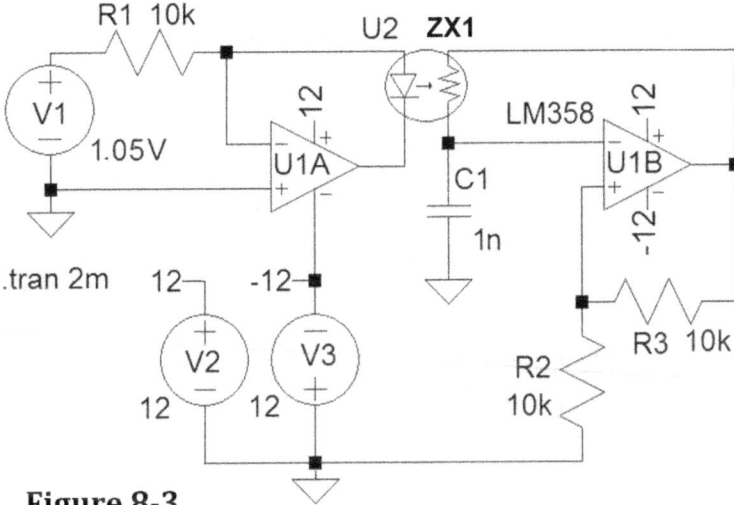

Figure 8-3

Theoretical calculations:

$$T = 2C1R_{ZX1} \ln\left(1 + \frac{2R2}{R3}\right). \qquad f_0 = \frac{1}{T}.$$

$$T = 2(1 \times 10^{-9})(100k)\ln\left(1 + \frac{2(10k)}{(10k)}\right) = 220\mu S. \qquad f_0 = \frac{1}{T} = \frac{1}{220\mu S} = 4545Hz.$$

The simulation result below shows a period of 220µS which corresponds to a frequency of 3509Hz. The longer period is due to the op-amp's slew rate.

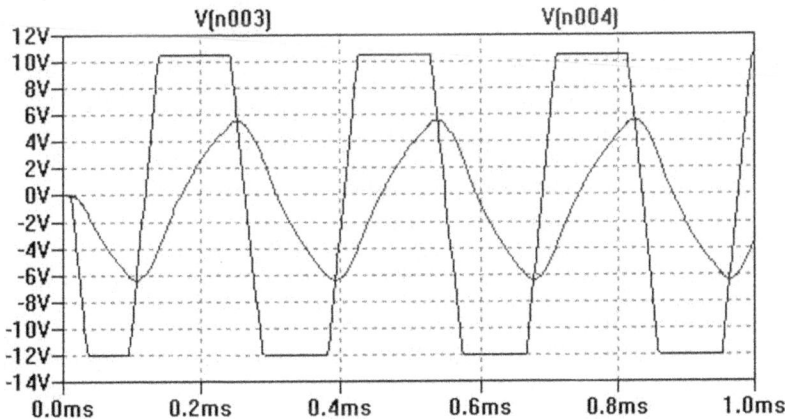

Project 15: Audio AGC and Compressor Circuits

AGC circuits are used in receiver circuits (radio, TV, communications) to maintain a relatively constant output signal level. Compressors are used to reduce the dynamic range of the input signal. AGC and compressor circuits are basically similar. AGC circuits usually control the voltage gain of an RF (radio frequency) amplifier while compressor circuits usually control the voltage gain of an audio amplifier.

The functional block diagram of a voltage controlled amplifier in an AGC or compressor is shown below. The output signal is rectified and filtered to provide a DC control voltage for the amplifier. An increase in signal amplitude reduces the voltage gain of the amplifier.

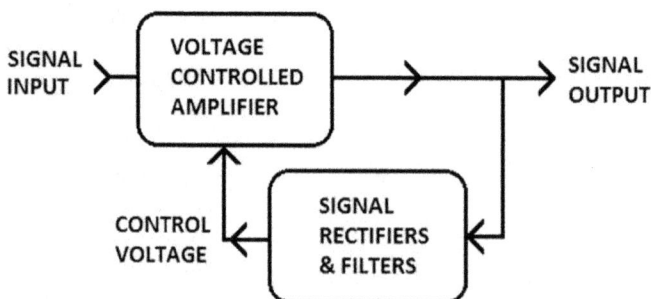

This focus of this project is audio compression. Audio compressors are widely used in the music industry for microphones, musical instruments, and audio mixers. There are three common implementations usually designated as VCA, FET, and photo-resistor. VCA refers to a voltage controlled amplifier of the type where the gain is controlled by a transistorized integrated circuit. FET type refers to gain control using the transconductance properties of an FET. Photo-resistor type usually refers to gain control using a photo-resistor opto-coupler, such as the NSL-32.

Although the VCA is considered the most modern of the three methods, FET and photo-resistor compressors are still widely used. Some say that the photo-resistor type is somehow more musically pleasing. The main problem with the photo-resistor type is its slow response time, typically several milliseconds rise time and tens to hundreds of milliseconds fall time.

In compressor lingo, rise time is similar to attack time and fall time is similar to release time. The photo-resistor type sets a limit on the shortest attack and release times possible. The NSL-32SR3 is now available as an improved version of the NSL-32 with considerably shorter rise and fall times. Refer to the appendix for details.

Simulating a Photo-Resistor Compressor Circuit

Simulations using the ZX1 take a long time to run due to the RC network in the model which simulates rise and fall times. A model without the RC network, the ZX1A, is used here. Refer to the appendix for more information. An LT1366 rail-to-rail op-amp can be substituted for the LM358.

Figure 8-4 below shows the LTspice circuit of a voltage controlled amplifier whose gain is controlled by the voltage on pin 5 of U1B, Vc. Vo is rectified by the schottky diodes, D1 and D2, filtered by R3, R5, C1, and output as Vc.

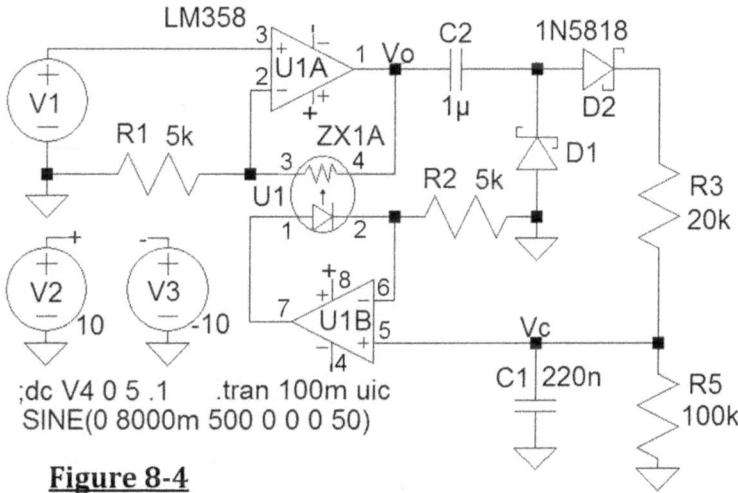

Figure 8-4

The characteristics of the amplifier and rectifier circuits will be measured in the "open loop" mode. Vc is disconnected from U1B, pin 5. A voltage source, V4, is added and connected to U1B pin 5. V4 supplies the control voltage. V1 is set to 100mV DC and V4 is swept from 0 to 5V in 0.01 volt increments. The output voltage, Vo, as a function of the control voltage, V4, is shown below.

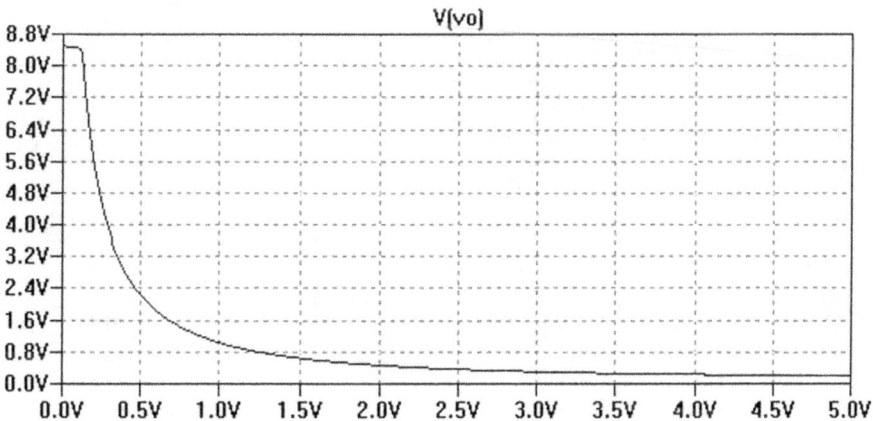

101

Next, V1 is set to produce a 500Hz, 100mV peak amplitude, sine wave (10 cycles). Analysis type is transient with a stop time of 10mS. V4 is set to the values in the table on the right.

The peak output voltage, Vo, and the DC control voltage, Vc, is measured for each value of V4. Refer to the table on the right.

OPEN LOOP	V1 = 100mV		
V4	Vo mVp	Vc	Gain
0.125	8000	8.6	80
0.25	4500	4.6	45
0.5	2260	2.1	22.6
1	1050	0.8	10.5
2	470	0.31	4.7
4	230	0.13	2.3
8	170	0.09	1.7

The simulation result for V4 equal to 0.5 volts is shown below. Note the ripple on Vc. The ripple can be reduced by increasing the value of C1 at the expense of increasing the rise time.

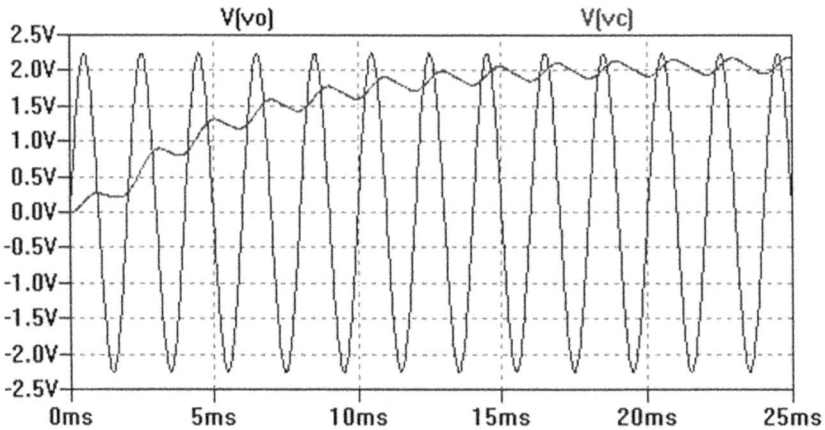

Next V4 is removed and Vc is connected to U1B, pin 5 for closed loop analysis. V1 is set to produce a 500Hz, 100mV peak amplitude, sine wave (50 cycles). Analysis type is transient with a stop time of 100ms.

Vo and Vc are measured. The measurements are repeated for V1 values of 250mV, 500mV, 1000mV, 2000mV, and 4000mV.

Simulation results are presented in the table on the right.

CLOSED LOOP				
V1 mVp	Vo mVp	Vc V	I_{LED} µA	Gain
100	1100	0.92	184	11
250	1690	1.5	300	6.7
500	2250	2.1	420	4.4
1000	3000	3	600	3
2000	4400	4.4	880	2.2
4000	6800	7.3	1460	1.7

The results show that the amplifier's voltage gain is 11 for a 100mV peak amplitude input signal, but only 1.7 for a 4000mV peak amplitude input signal. This is a compression ratio of 6.45 to 1.

In figure 8-5 below a voltage controlled switch is used change the amplitude of the input signal 20ms after the start of the simulation. The switch is set to be on at the start of the simulation. It will turn off after one time constant of R7 and C3 (1uF X 20k = 20ms).

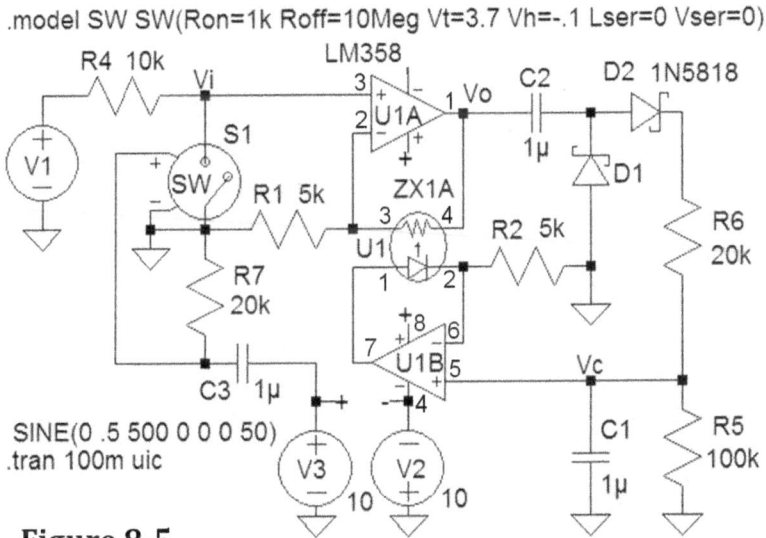

Figure 8-5

The on resistance of the switch is 1000 ohms. When the switch is on the input amplitude, Vi = 0.91 X V1. When the switch is off, Vi =V1, an increase of 11 times. V1 is set to a 500mV peak, 500Hz, sine wave. At the start of the simulation Vi is 45.5mV peak. After 20ms Vi increases to 500mV peak. Refer to the graph below.

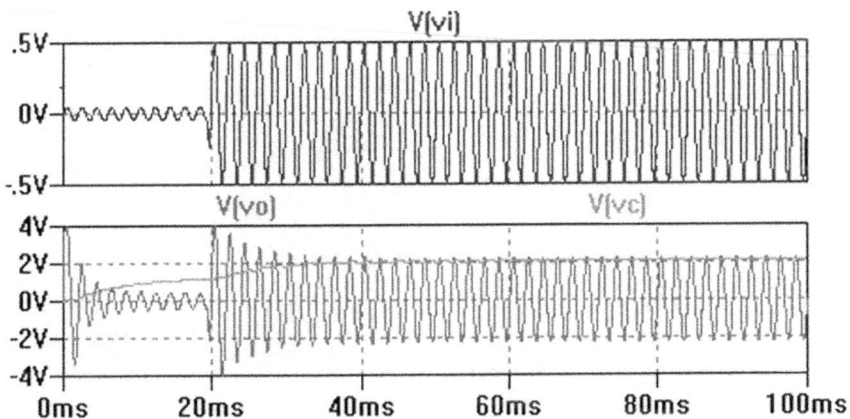

103

The graph above shows that when the input amplitude, V(vi), increases to 500mV peak, the output amplitude, V(vo), increases to 4V peak and decreases to 2V peak in about 15ms. The control voltage, V(vc) is shown climbing from 0 volts to 2VDC. Note that the input amplitude increased by a factor of 11, but the steady state output amplitude, V(vo), only increased by a factor of 4. This is a compression factor of 2.75 or 8.8dB.

Experiment Suggestions

1. Simulate the circuit with R1 equal to 3k. This increases the gain of the amplifier. Compare the compression ratio to when R1 equals 5k.

2. Reset R1 to 5k. Simulate the circuit with R2 equal to 10k. Compare the compression ratio to when R2 equals 5k.

3. Vary the values of R6 and C1 to vary the attack and decay time.

Project 16: Tremolo Circuits

"Tremolo" usually refers to a rapid variation of amplitude of a musical tone. It is also used to refer to a rapid change in pitch of a musical tone, although this effect is usually referred to as "vibrato".

This project uses a photo-resistor opto-coupler to vary the amplitude of the input signal at a rate of about 1Hz. Typically the rate is adjustable between about 1Hz and about 20 Hz. A block diagram of the circuit is shown below.

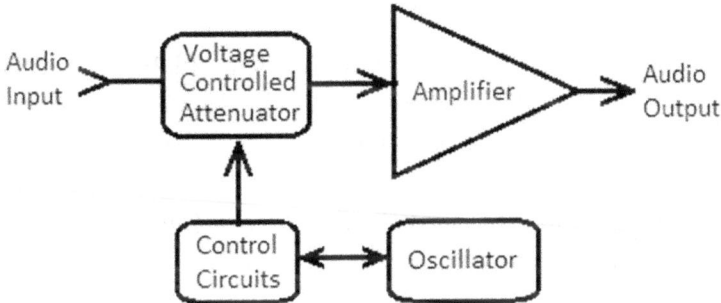

This project demonstrates the basic concept of the tremolo. A simple relaxation oscillator circuit is used to produce a waveform which is used to control the signal amplitude. Tremolos may also have controls to vary the oscillator wave form shape, for example, triangle, sine, and square shapes.

Simulation is done using *TINA-TI*. This is a free version of *Design Soft's TINA* which is available as a free download from *Texas Instruments*. TINA-TI has no limit on the number of nodes or parts and is just as useful as LTspice. The simulation may also be done using *LTspice*; however *LTspice* seems to be considerably slower performing the transient simulations required for his project.

Both *LTspice* and *TINA-TI* are excellent free simulators. Almost all simulators available today are based on Berkley SPICE (**S**imulation **P**rogram with **I**ntegrated **C**ircuit **E**mphasis). It originated as a command line program 1972 from the EECS Department of the University of California at Berkley. Manufacturers may optimize a particular version of SPICE for their ICs. For example, *LTspice* is optimized for switching regulator circuits.

This *TINA-TI* simulation requires a *TINA-TI* compatible photo-resistor opto-coupler model. Please refer to the appendix, "ZX4 Photo-Resistor Opto-Coupler *TINA-TI* Model". The ZX4 model's characteristics are identical to the ZX1A *LTspice* model and therefore it is similar to the NSL-32.

The TINA-TI schematic of the Tremolo circuit is shown below in figure 8-6. R5, R9, and the ZX4 photo-resistor form a voltage controlled attenuator (voltage divider). Pin 6 of the ZX4 is the LED anode, pin 5 is the cathode. The controlled resistance is between pins 3 and 0.

U3 is a relaxation oscillator (refer to chapter 4 and figure 4-1). Its frequency can be varied with R13 and C4. When C4 is 10μF and R13 is 100k, the frequency is about 0.7Hz. Changing R13 to 10k changes the frequency to about 7Hz. R13 could be a 100k ohm potentiometer. The oscillator's output, the voltage waveform across C4, is buffered by U4.

U4/2 is a voltage controlled current source. The oscillator signal is ac coupled to the current source by R14 and C4. Diodes D1 and D2 shape the waveform and clip it to an amplitude of about 1V peak-to-peak. Diode D3 adds a positive offset of about 0.5 volts. R8 controls the opto-couplers LED current. A value of 1k results in a current of 1mA per volt. R8 controls the "depth" of the tremolo effect. It could be a potentiometer.

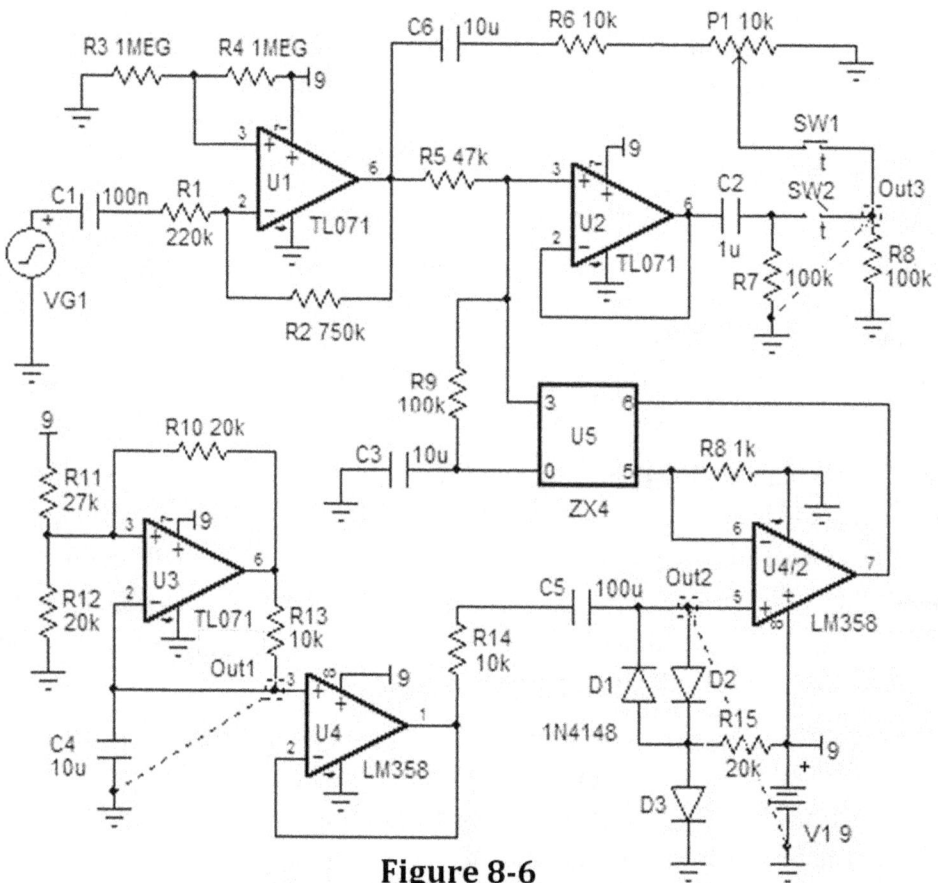

Figure 8-6

106

U1 is signal pre-amplifier with a voltage gain of 3 and an input impedance of 220k ohms. Its output is applied to the controlled voltage divider and to the by-pass network (C6, R6, and P1). U2 is a buffer for the controlled voltage divider.

SW1 and SW2 are TINA's time controlled switches which switch the tremolo's output between by-pass and tremolo. P1 sets the non-tremolo output level. SW1 and SW2 would normally be an SPDT switch (maybe a foot switch).

Simulation

VG1 is set to produce a 50Hz, 200mV peak amplitude, sine wave. Outputs are inserted to monitor the oscillator (out1), LED current (out2), and signal output, (out3). Transient simulation is performed with a stop time of 20 seconds. SW1 is set to be on for 0 to 10 seconds and off for 10 to 20 seconds. SW2 is set to be off for 0 to 11 seconds and on for 11 to 20 seconds. The results are presented below.

Out1 shows the oscillator output of about 2V$_{p-p}$ as a 0.7Hz triangle wave.

Out2 shows that the range of the control voltage on pin 5 of U4/2 is between 100mV and 1007mV. This corresponds to a change in LED current from 0.1mA to 1.07mA.

The resistance, Rx, versus LED current curve for the ZX4 shows that this corresponds to a change in resistance from about 101k to 4.3k.

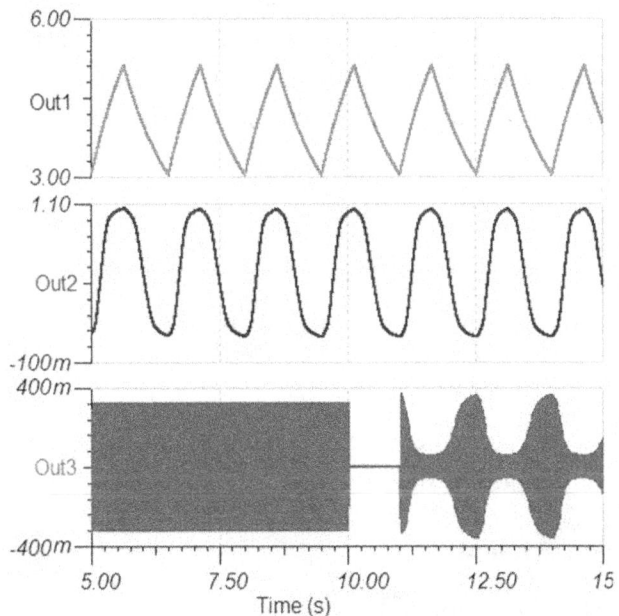

Out3 shows the output between 5 seconds and 15 seconds of the transient simulation. From 0 to 10 seconds the output is the input signal without the tremolo effect. The amplitude of this signal can be adjusted with potentiometer P1. At t = 10 seconds SW1 opens and no signal is output. At t = 11 seconds SW2 closes and the output is the input signal with the tremolo effect.

Out3 shows that the tremolo effect is varying the output amplitude between 100mV peak-to-peak and 714mV peak-to-peak.

Changing R13 to 10k increases the oscillator frequency to about 7Hz. Output 3 below shows the amplitude variation of the output as well as the individual cycles of the 50Hz input signal.

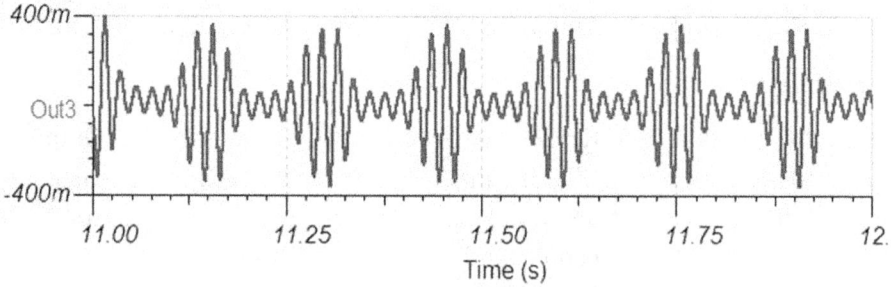

Theoretical

The minimum and maximum output amplitudes can be calculated using the values of the opto-coupler resistance, Rx. Minimum output, V_{min}, occurs when Rx equals 4.3k and maximum output, V_{max}, occurs when Rx equals 101k. First the resistance of Rx in parallel with 100k is calculated.

$$Rx_{MIN} = \frac{(4.3k)(100k)}{4.3k+100k} = 4.12k \qquad Rx_{MAX} = \frac{(101k)(100k)}{101k+100k} = 50.2k$$

Next the minimum and maximum amplitudes are calculated.

$$V_{min} = \frac{4.12k(1.36V_{p-p})}{47k+4.12k} = 109mV_{p-p} \qquad V_{max} = \frac{50.2k(1.36V_{p-p})}{47k+50.2k} = 702mV_{p-p}$$

As expected, calculated results are in approximate agreement with the simulation results.

Experiment

The simulation of this circuit could be done with LTspice using the ZX1A opto-coupler model.

This circuit may be may be built, tested, and optimized on a breadboard. The optimized circuit could then be built on a perf-board or pc-board as a tremolo petal project. The results of a breadboard experiment are presented on the next page.

NSL-32 Test and Measurement

The resistance versus current characteristics of the NSL-32 used in this experiment was obtained using the voltage controlled current source circuit below. Photo-resistor resistance is measured for selected LED currents between 0.1mA and 4ma. The results are presented in the table.

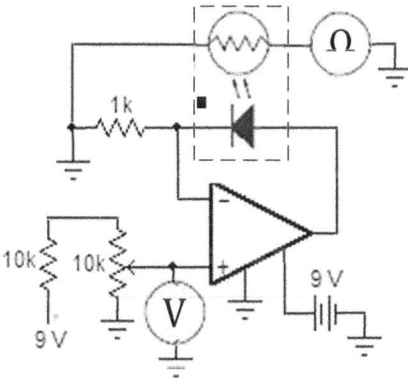

V Volts	Current mA	Resistance Ω
0.1	0.1	78k
0.2	0.2	27k
0.4	0.4	10k
2	1	3.4k
2	2	1.6k
4	4	0.89k

A voltage controlled current source can be used to measure the resistance vs current characteristics of an opto-coupler.

The circuit of figure 8-6 was built on a breadboard using the part values given except for R13 and C4, which were set to 100k and 6uF respectively. These determine the tremolo frequency.

The depth of the tremolo effect is determined by the value of R8. Three values of R8 were used: 1k, 2.2k, and 4.7k. R8 could be replaced by a 470 ohm resistor and a 5k ohm or 10k ohm potentiometer. This potentiometer would be the tremolo's depth control.

A 20Hz sine wave is applied to the input and its amplitude is adjusted so the there is 1.0V$_{p-p}$ at the output of U1. The tremolo period is 0.5 seconds corresponding to a frequency of 2Hz.

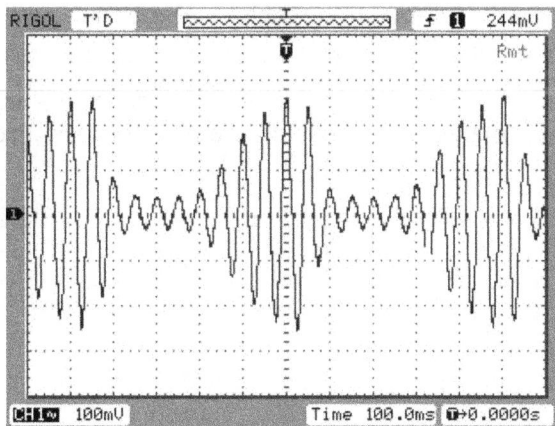

V$_{MIN}$ = 80mV$_{p-p}$.

V$_{MAX}$ = 500mV$_{p-p}$.

R8 = 1k.

V_{MIN} = 140mVp-p.

V_{MAX} = 560mVp-p.

R8 = 2.2k.

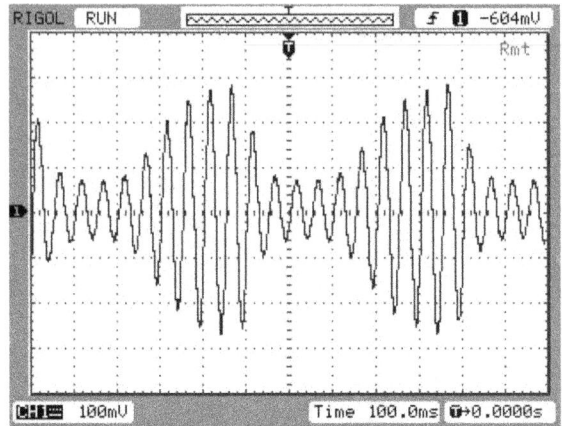

V_{MIN} = 340mVp-p.

V_{MAX} = 600mVp-p.

R8 = 4.7k.

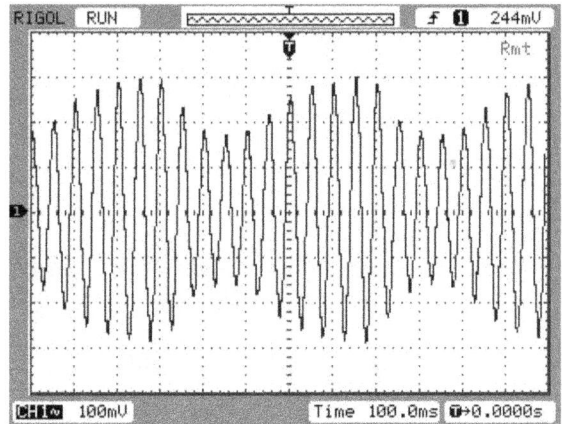

Experiment Suggestions

1. Simulate the tremolo circuit with LTspice using the ZX1A macro.

2. Build the circuit on a breadboard. Compare your breadboard results to your simulation results. Note that published photo-resistor opto-coupler data show that the resistance versus current curve for any particular part number can vary considerably, especially for currents below 100µA. This makes the design of a tremolo for mass production more difficult.

Appendix

Op-Amp Data Table

Part #	Price	SR V/μs	GBP MHz	Bias nA	Ios nA	Vos mV	10Hz Vn√Hz	1KHz Vn√Hz	1KHz In√Hz
L272	0.66	1	350	300	50	15	(power opamp)		200pA
LM358	0.49	0.3	0.7	45	5	3		40nV	.1pA
LM741	0.71	0.5	1.5	80	20	2		30nV	.15pA
LT1001	3.32	0.3	0.8	0.7	0.4	0.02	10nV	11nV	3pA
LT1013	2.17	0.4	1	12	0.2	0.06		24nV	.07pA
LT1055	3.99	12	5	0.03	0.01	0.1	30nV	15nV	2fA
LT1112	4.72	0.3	0.75	0.08	0.06	0.025	16nV	14nV	.03pA
LT1113	6.03	4	5.6	0.32	0.035	0.5	17nV	4.5nV	.01pA
LT1128	4.97	5	20	30	18	0.02	1nV	1.1nV	1pA
MCP6271	0.38	0.9	2	0.001	0.001	1		20nV	3fA
NE5532	0.94	9	10	200	10	0.5	10nV	5nV	2.7pA
OP07	0.91	0.3	0.6	2	1	0.06		12nV	.35pA
OP27	3.31	2.8	8	15	12	0.03	3.2nV	3.8nV	2pA
OP270	6.13	2.4	5	10	3	0.05	3.6nV	3.2nV	.6pA
OPA228	3.33	11	33	3	3	0.01	3.5nV	3nV	.4pA

SS495 Hall Sensor

		SS495A	SS495A1	SS495A2
Supply	VDC	4.5 to 10.5	4.5 to 10.5	4.5 to 10.5
Supply I @ 25°C	Typ.	7mA	7mA	7mA
Output I Typ. Source	Vs>4.5V	1.5mA	1.5mA	1.5mA
Magnetic Range	Typ.	−670 to +670	−670 to +670	−670 to +670
Output V Span	Typ.	0.2 to (Vs -0.2)		
Null @ 0 Gauss, V	Typ.	2.50 ± 0.075	2.50 ± 0.075	2.50 ± 0.100
Sensitivity (mV/G)	Typ.	3.125 ± 0.125	3.125 ± 0.094	3.125 ± 0.156
Linearity, % of Span	Typ.	−1.0%	−1.0%	−1.0%

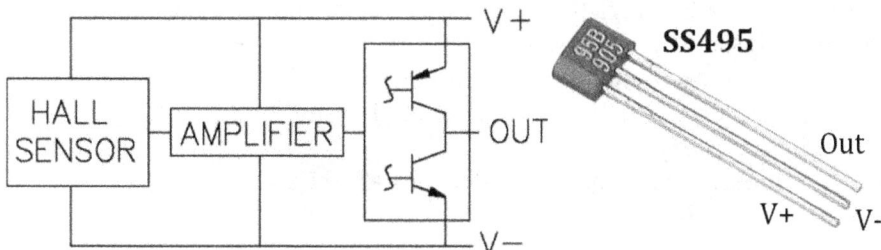

LM385-1.2 Voltage Reference

Reverse Breakdown Voltage (I =20 µA to 20 mA)
1.192V - 1.235V - 1.27V
Minimum Operating Current, 20µA
Reverse Breakdown Voltage Change with Current
20µA to 1.0 mA: 1.5mV
1.0 mA to 20 mA: 25mV

N.C.
Cathode
Anode

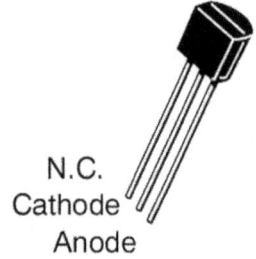

TIP120 NPN Power Transistor

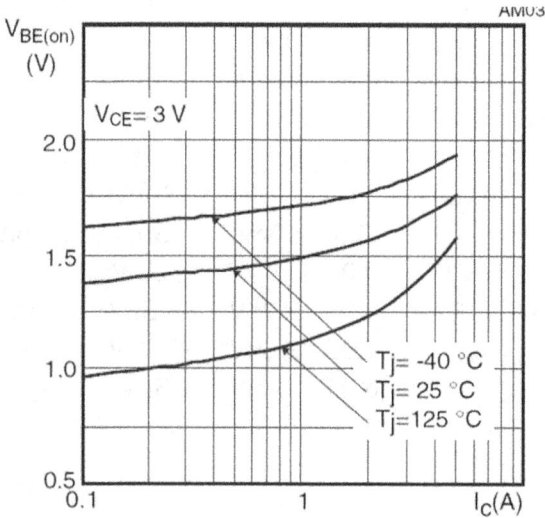

AMO3.

$V_{BE(on)}$ (V)

V_{CE}= 3 V

2.0

1.5

1.0

Tj= -40 °C
Tj= 25 °C
Tj=125 °C

0.5
0.1   1 I_C(A)

Hfe = 1000
Rthj-case Thermal resistance
junction-case max. 1.92 °C/W

Rthj-amb Thermal resistance
junction-ambient max. 62.5°C/W

NPN TIP120, PNP TIP125: V_{CBO} Collector-base (I_E = 0) 60V.
V_{CEO} Collector-emitter (I_B = 0) 60. V_{EBO} Emitter-base (IC = 0) 5 V. $I_{C(MAX)}$: 5A.

RTC – 4.5-395 – Vertical Geophone

•Natural Frequency (fn): 4.5Hz, •Tolerance: +/- 0.5Hz
•Max. tilt angle for specified fn: 10 degrees
•Typical spurious frequency: > 160Hz
•Distortion: <= 0.2%
•Coil Resistance: 395ohm, Tolerance: +/- 5%
• "G" Sensitivity: 22.0 V/m/s (0.698 V/in/s), Tolerance: +/- 7.5%
• "m" Moving Mass: 11.2 g (0.395 oz)
• "z" Maximum coil excursion p.p.: 1.52 mm (0.060 in)
•Diameter: 25.4 mm (1 in), Height: 33.0 mm (1.26 in)
•Weight: 86 g (3.03 oz)
•Operating temperature range: -45 Celcius to +100 Celcius

112

NJL7502L Photo-Transistor

The NJL7502L is a photo transistor whose spectral response is similar to the human eye.

PARAMETER	Ta=25°C TEST CONDITION	MIN	TYP	MAX
Photocurrent	V_{CE}=5V, White LED, 100Lux	15 µA	33 µA	73 µA
Dark I I_D	V_{CE}=20V			0.1 µA
Peak λ $λ_P$			560nm	
Half Angle $θ_{1/2}$			±20°	
Rise Time	V_{CE}=5V. I_C=1mA, R_L=100Ω		10µs	
Fall Time	V_{CE}=5V. I_C=1mA, R_L=100Ω		10µs	

Long lead = collector

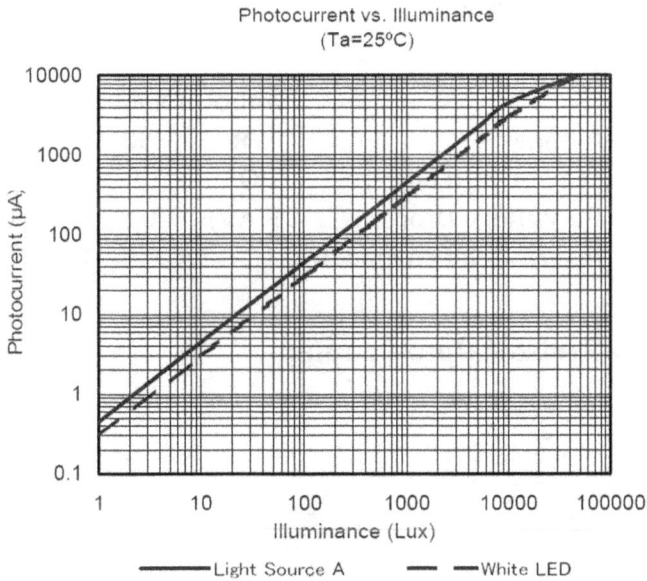

Photocurrent vs. Illuminance
(Ta=25ºC)

————Light Source A   — —White LED

Resistor Color Code

1 Brown   6 Blue
2 Red   7 Violet
3 Orange 8 Grey
4 Yellow   9 White
5 Green

1% Tolerance

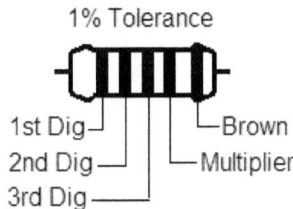

1st Dig┘
2nd Dig——
3rd Dig———
—Brown
—Multiplier

5% Tolerance

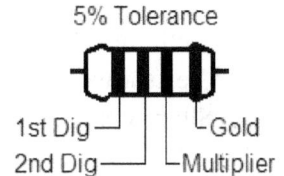

1st Dig—
2nd Dig———
└Gold
└Multiplier

SD103A Schottky Diode

VISHAY SD103A	Schottky Diode					
PARAMETER	TEST COND.	PART	SYMBOL	MIN.	MAX.	UNIT
Rev. bkdn volt.	IR = 50µA	SD103A	V(BR)	40		V
Leakage current	VR = 30 V	SD103A	IR		5	µA
Forward voltage	IF = 20 mA		VF		370	mV

1N4148 Signal/Switching Diode

VISHAY 1N4148	Signal/Switching Diode				
PARAMETER	TEST COND.	SYMBOL	MIN.	MAX.	UNIT
Forward voltage	IF = 10 mA	VF		1000	mV
Reverse current	VR = 20 V	IR		25	nA
Breakdown voltage		V(BR)	100		V

1N400X Rectifier Diodes

Rectifier Diodes					
Rating	Symbol	1N4001	1N4002	1N4003	Unit
Peak Reverse Voltage	VR	50	100	200	V
†Average Rectified Forward Current	IO	1	1	1	A
Average Forward Voltage Drop,	VF(AV)	0.8	0.8	0.8	V
(IO = 1.0 Amp, TL = 75°C, 1 inch leads)					

NSL-32 Photo-Resistor Opto-Coupler

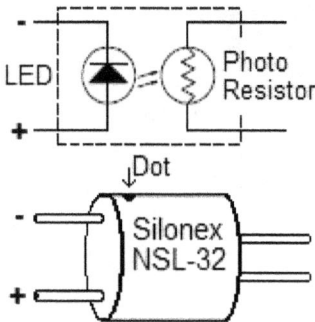

A graph of typical LED current and resistance is on the right for the Silonex NSL-32 Opto-coupler.

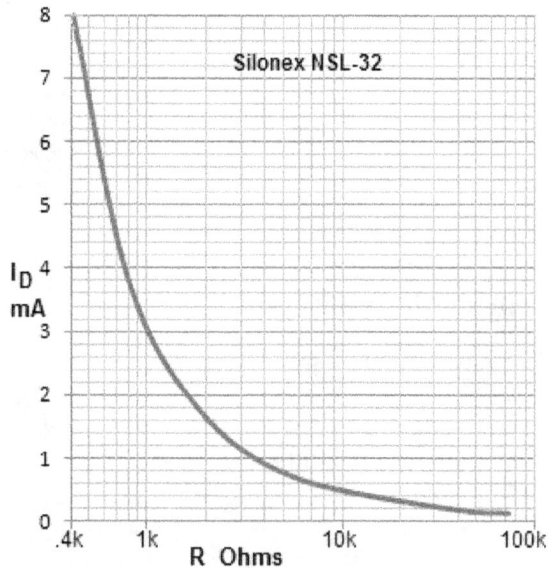

Photo-Resistor Output Opto-Couplers   *Advanced Photonix, Inc*

Part	I(F) mA	Max V	P(D) mW	On Res	T(R) ms	T(F) ms
NSL-28	40	100	50	100k (max)	5	80
NSL-32	40	60	50	1k	3.5	500 (max)
NSL-32SR3	25	60	50	60	5	10
NSL-32H-103	40	60	50	1.65k (max)	3.5	500 (max)
NSL-32SR2	25	60	50	40	5	80
NSL-32SR2S	25	60	50	40	5	80

ZX1 Photo-Resistor Opto-Coupler LTspice Model

This model has a built in rise time of about 5ms and fall and fall time of about 120ms. These simulated rise and fall times can actually make some transient simulations too slow. The ZX1A which does not have simulated rise and fall times built in may be preferable for some circuits.

The file below goes into the */LTspice* /lib/sub directory as "ZX1.sub". It may be necessary to start *LTspice* as an administrator to do this.

```
** ZX1 Photo-Resistor Opto-Coupler LTspice Model **
** Similar to the NSL-32 **
** www.zapstudio.com      Sid Antoch 2015 **

** Connect:   + - R R
.subckt ZX1   1 2 3 4
VS 10 4 0
R1 9 10 1m
R2 8 2 20k
H2 7 2 Value={exp(exp(exp(exp(-1*(((I(V1)*40)**(.4)))))))}
V1 6 2 0
D1 1 6 AOT-2015
H1 3 9 VALUE={I(VS)*V(5,0)}
R4 5 7 100k
C1 5 2 100n
W1 5 8 H2 CSW

.model D D
.lib C:\Program Files (x86)\LTC\LTspiceIV\lib\cmp\standard.dio
.model CSW CSW(Ron=.1 Roff=10Meg It=1u Ih=0)
.ends.model  D  D
.lib C:\Program Files (x86)\LTC\LTspiceIV\lib\cmp\standard.dio
.ends
*Note: the .lib paths may need to be changed for some pcs.
```

ZX1A Photo-Resistor Opto-Coupler LTspice Model

The file below goes into the LTspice /lib/sub directory as "ZX1A.sub". It may be necessary to start LTspice as an administrator to do this.

```
** ZX1A Photo-Resistor Opto-Coupler LTspice Model **
** Similar to NSL-32 with much faster rise and fall times **
** www.zapstudio.com      Sid Antoch 2015 **

** Connect:   + - R R
.subckt ZX1A 1 2 3 4
VS 10 4 0
R1 9 10 1m
R2 5 2 20k
H2 5 2 Value={exp(exp(exp(exp(-1*(((I(V1)*40)**(.4)))))))}
V1 6 2 0
D1 1 6 AOT-2015
H1 3 9 VALUE={I(VS)*V(5,0)}
.model D D
.lib C:\Program Files (x86)\LTC\LTspiceIV\lib\cmp\standard.dio
.ends
*Note: the .lib paths may need to be changed for some pcs.
```

Add the part to the subcircuit directory:

Carefully type the subcircuit file into a text editor. Save the file with a ".sub" extension into *LTspice's* sub circuit directory. The sub-circuit must be associated with a symbol. An easy way to do this is to "AutoGenerate" it.

1. Go the *LTspice* subcircuit directory and open the desired sub circuit file into *LTspice*. You will need to select all file types (*.*) to see the files in the directory. In the example below, the file "ZX1A.sub" was opened.

2. Right-click the line with the subcircuit name, and select *Create Symbol*:

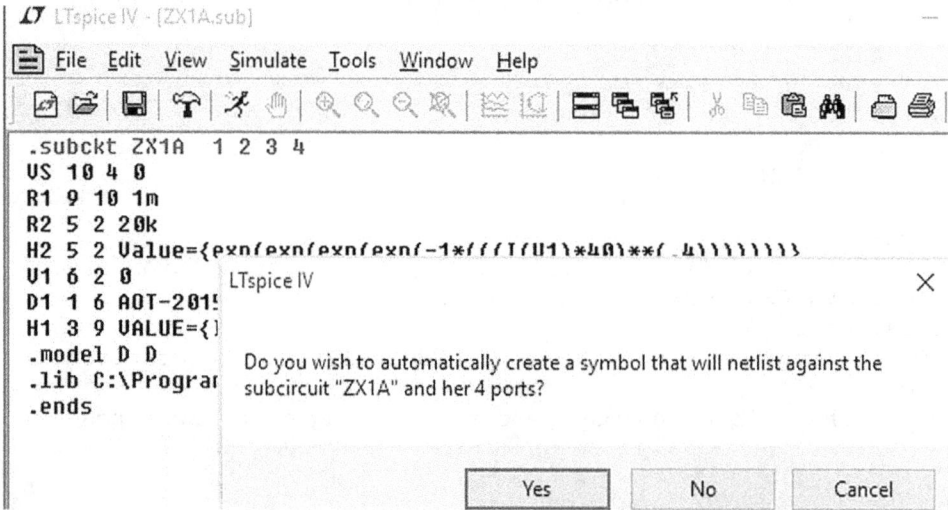

Select yes. The "AutoGenerated" [ZX1A.asy] file is shown on the right.

The symbol can be edited. For example, an LED symbol could go between pin 1 (anode) and pin 2 (cathode). A resistor symbol could go between pin 3 and pin 4.

This symbol will be saved in the AutoGenerated directory of the component library. To place it in a schematic, select it from the AutoGenerated directory.

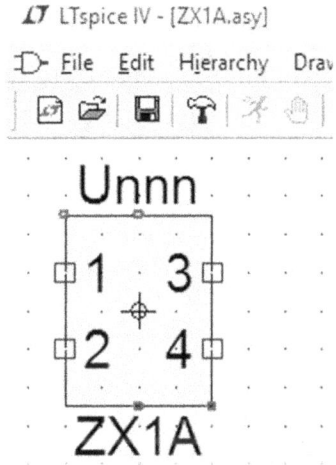

It is important that the "ModelFile" attribute in the Symbol Attribute Editor has the correct path to the ZX1A.sub file. ZX1A.sub may be sufficient. The full path is typically: C:\Program Files (x86)\LTC\LTspiceIV\lib\sub\ZX1A.sub.

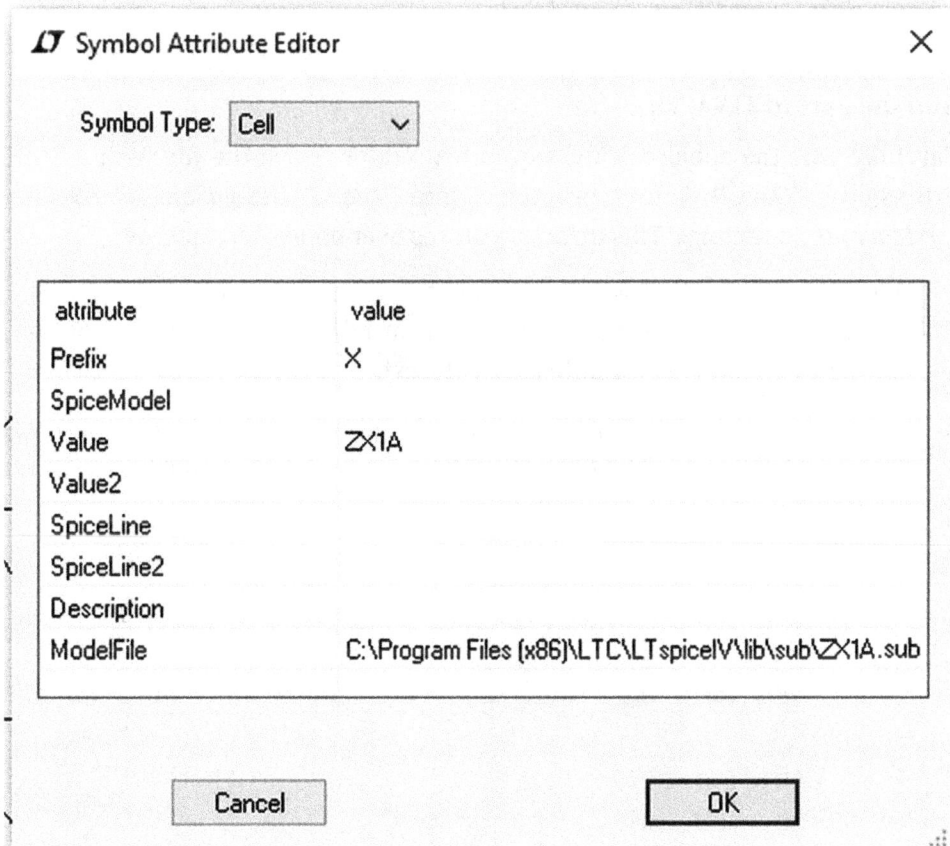

LT Symbol Attribute Editor   ✕

Symbol Type: Cell ⌄

attribute	value
Prefix	X
SpiceModel	
Value	ZX1A
Value2	
SpiceLine	
SpiceLine2	
Description	
ModelFile	C:\Program Files (x86)\LTC\LTspiceIV\lib\sub\ZX1A.sub

Cancel   OK

ZX4 Photo-Resistor Opto-Coupler *TINA-TI* Model

This Photo-Resistor Opto-Coupler model was made using *TINA-TI*. It has the same characteristics as the ZX1A model developed for LTspice and is therefore similar to the NSL-32.

```
** ZX4 Macro for NSL-32 Photo-Resistor Opto-Coupler Equivalent
** www.zapstudio.com        Sid Antoch 2015
** Connect: + - R R
.subckt ZX4 6 5 3 0
ECS2          3 2 VALUE = {I(VCS2_V1)*V(1,0)}
VCS2_V1       2 0
ECS1          1 0 VALUE = {EXP(EXP(EXP(EXP(-
1*(((I(VCS1_V1)*40)**(.4)))))))}
VCS1_V1       4 5
R2            1 0 1K
DLED1         6 4   D_CQX35A_1

.MODEL D_CQX35A_1 D( IS=5.62P N=2.8 BV=5 IBV=100U RS=420M
+       CJO=35P VJ=750M M=330M FC=500M TT=100N
+       EG=1.11 XTI=3 KF=0 AF=1 )
.END
```

Add the part to TINA-TI:

Carefully type the subcircuit file into a text editor. Save the file with a ".CIR" extension, "ZX4.CIR" for example, into the "*DesignSoft\{Tina9-TI-...-...}\Macrolib*" directory. This directory stores user generated macros.

1. Open TINA-TI and click on *Tools* in the main menu. Select "New Macro Wizard". Enter macro name, select "From File" (maybe the only option), and select file the file (ZX4.CIR). Select Next>.

2. Select Auto generate shape. A rectangle is generated with pin numbers. Pin 6 is the diode anode and pin 5 is the cathode. The resistance is between pin 3 and pin 0.

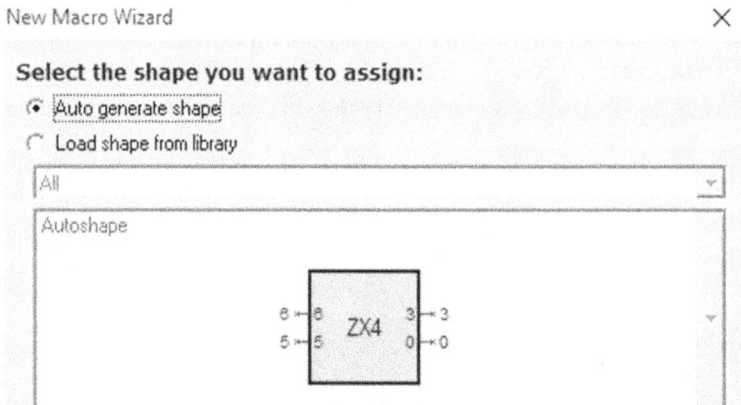

3. Click on *Next*. The new macro can be placed on the schematic and saved in the "macrolib" directory.

4. To use the macro in a schematic, click on "*Insert*" in the main menu and select "*Macro..*". The macro should be in the TINA-TI directory, for example: \Documents\DesignSoft\{Tina9-TI-...-...}.

Resources

Parts
www.mouser.com
www.digikey.com
www.jameco.com
www.newark.com
www.alliedelec.com
www.allelectronics.com

Equipment
www.agilent.com
www.tek.com
www.rigolna.com
www.siglent.com

USB Instruments
www.easysync-ltd.com
www.picotech.com

Simulation Software
LTspice from Linear Technology: www.linear.com
TINA-TI from Texas Instruments: www.ti.com

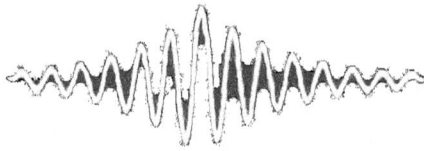